Walks in the Wild

By the same author:

The Hidden Life of Trees
The Inner Life of Animals
The Weather Detective
The Secret Network of Nature

Walks in the Wild

A guide through the forest

PETER WOHLLEBEN

Translated by Ruth Ahmedzai Kemp

RIDER

LONDON SYDNEY AUCKLAND JOHANNESBURG

1 3 5 7 9 10 8 6 4 2

Rider, an imprint of Ebury Publishing,
20 Vauxhall Bridge Road,
London SW1V 2SA

Rider is part of the Penguin Random House group of companies whose
addresses can be found at global.penguinrandomhouse.com

Copyright © Piper Taschenbuch 2017

First published in Germany as *Gebrauchsanweisung für den Wald* in 2017

First published by Rider in 2019

www.penguin.co.uk

A CIP catalogue record for this book is available from the British Library

ISBN 9781846045578

Typeset in 12/15.5pt Bell MT
by Integra Software Services Pvt. Ltd, Pondicherry

Printed and bound in Great Britain by Clays Ltd, Elcograf S.p.A.

Penguin Random House is committed to a sustainable future for
our business, our readers and our planet. This book is made
from Forest Stewardship Council® certified paper.

Contents

Introduction   vii

1. A Walk in the Wild   1

2. The Woodland Detective   13

3. Spotting Wild Animals   23

4. The Right to Roam and Pick Your Own   33

5. Freshly Washed and Bitten   41

6. A Ticking Time-Bomb   49

7. Good Day and Good Hunting   59

8. The Fox: A Friendly Foe   75

9. Little Red Riding Hood Sends Her Regards   85

10. Trees – As You've Never Known Them Before   103

11. Is It Really Love?   117

12. A Pocket Dictionary   127

13. The Lumberjack Lads   133

14. Conservation: Love with an Impact   143

15. The Woods in Bad Weather   153

16. Litterbugs and Forest Fires   159

17. Lost Without a Watch or a Compass?   163

18. Forest Survival Skills   169

19. When the Forester Becomes an Undertaker   185

20. Is that Allowed?   199

21. The Forest at Night   205

22. Dress Code   215

23. The Forest at Home   221

24. Woodland Walks in February   229

25. Woodland Walks in May   233

26. Woodland Walks in August   239

27. Woodland Walks in November   243

28. Forest School: Child-Friendly Adventures
in the Woods   247

Conclusion   257

Notes   259

Introduction

WHEN MY PUBLISHER ASKED ME if I would like to write an instruction manual for the forest, I leapt at the opportunity. I love forests and they have determined the course of most of my life. But I stumbled accidentally into working with woodlands. I intended to study biology, because – like so many school leavers – I didn't quite know what to do with my love of nature. My mother spotted a small advert in the paper: the local state forestry administration here in the Rhineland-Palatinate was seeking applicants for a study programme. I applied, was accepted and spent the next four years alternating between practical placements and lecture halls.

What I subsequently encountered in the field – or rather, in the woods – was not at all what I had imagined or hoped. The use of heavy machinery that destroyed the forest floor was just the tip of the iceberg. Poisonous insecticides, clear cutting, and felling the oldest trees (mature beeches, which I love so much) – these were all part and parcel of an approach to forestry that left me feeling increasingly alienated. During my studies, I had been taught that all of this was key to maintaining healthy forests. As strange as it may seem to you and me, this is what thousands of students are led to believe by their lecturers. Initially uncomfortable about working in such a way, I gradually began to reject this approach. I was unsure how I could pursue a lasting career when battling against an approach that I disagreed with so deeply.

Then, in 1991, I came across the publicly owned ancient beech forest at Hümmel, in Germany's mountainous Eifel region, which the municipality wanted to manage in an ecological and sustainable manner. Part of the forest is over 4,000 years old, making it one of the few untouched areas of land that still exist in Germany. Working together, we've succeeded in combining set-aside reserves with carefully tended plots for felling, while simutaneously encouraging the involvement of the local population. I started to run events for the public, ranging from forest survival skills training and building log cabins, to guided tours amidst the wonderful world of trees.

I was often asked by people who came on these walks where they could go and read more about what I was telling them. All I could do was shrug, because at that point I knew little of the literature on the subject. My wife urged me to write something down if only for the benefit of our visitors, so I sat down to pen a short guide while on holiday in Lapland. I sent the manuscript to several publishers and said to my wife, 'If no one accepts it by the end of the year, then writing isn't for me.'

As you can see, things turned out rather differently, and I have greatly enjoyed adding writing to the scope of my work. With my writing, I hope I can inspire many more people to develop a love of forests, because to my mind there are still far too few people exploring and making use of this natural asset. Not in the sense of the timber industry – no, in many respects there is already far too much tree felling going on. What I want to see more of are the sort of adventures, great and small, that are waiting for you amidst the trees while they are alive. And all you have to do is step into the woods and go for a walk.

1. A Walk in the Wild

I'M SURE YOU CAN PICTURE THE SCENE. You're out walking in the woods with children and sooner or later everything gets a bit lively. Perhaps they've caught something, or spotted something dart through the undergrowth, and they shriek with excitement; or perhaps it's just a noisy expression of sheer delight at being in the woods. The knee-jerk response of the adults in the group tends to be, 'Shhh! Not so loud!'

But why? Do deer really mind when people make a noise? Yes, wild animals like it when it's quiet, but not because they're sensitive to noise. When a storm howls through the treetops or a heavy downpour descends, the noise means they can no longer distinguish any other sounds. They wouldn't be able to pick out the footsteps of an approaching wolf or a lynx. An ability to hear is therefore a matter of life and death for a deer. For this reason, they are more comfortable in calm, dry weather conditions, where every step on a crackling twig is audible.

Noisy people in the woods don't cause too much of a nuisance to the animals because their sounds don't immediately fill the whole forest; the noise comes just from one direction. They also know that if we're being noisy, we're not on the prowl. For we humans are their biggest enemy when we come in the form of the hunters. Even if wolves and lynx are gradually finding their way back into some European landscapes, they will always be hugely outnumbered by their human cohabitants. It is no wonder that our wild animals' anxiety is mainly focussed on bipeds. When we walk along trails through the woods, joyfully singing or chatting away loudly, we send a reassuring signal to our fellow creatures that we are not out hunting. This is the case even with wildcats, which are extremely shy. Even they were hunted once because of their reputation as mean deer-stalkers. Wildcats catching deer? Really? The wildcat might be only a very distant relative of the domestic cat, but it isn't much greater in size. Can you imagine your tabby making a meal of a roe deer? Its small teeth are much too short, and its jaws can't open wide enough to get a grip on such a large animal. Nevertheless, for centuries the wildcat's deer-stalking prowess was considered common knowledge among hunters, meaning the feline predator was mercilessly tracked down. It's no wonder that they are very shy.

But the sounds of people strolling through the forest are not perceived as a danger by other species. One snowy January, I led a group of visitors through the old beech wood that I manage here at Hümmel. The hikers wanted to see our woodland burial ground. After we had looked around for an hour, we were heading back to the car park, when I realised I had left my backpack under a tree. Our work experience student offered to go back and fetch it for me. When he caught up with us a quarter of an hour later, he was buzzing. He had seen a wildcat saunter nonchalantly across the path. The animal had clearly waited close by until it was confident that the chatty humans had left the old forest. I had a similar experience a year later, on a hot July day, at the same car park for the woodland burial ground. I was talking to a colleague, leaning against my 4x4, when I saw a wildcat some fifty yards away from us, crossing the track from one stretch of woodland to another, quite untroubled by our presence. It didn't seem bothered by the road nearby either, suggesting that it was more likely to shy away from the presence of people quietly stalking through the undergrowth. From that, I'd conclude, there's no need to worry about making a racket in the woods, and least of all noisy children. Or rather, I wouldn't worry about disturbing the wild animals, though other adults might have something else to say about it!

A walk in the wild gives a sense of limitless freedom, yet I am often struck how that feeling of wilderness can vary from country to country. I love the great, wild landscapes of the south-western states of the USA, where you can walk for miles with no other human in sight. It's not because I seek to avoid other people, it's just the feeling these endless spaces inspire. In Europe, the vista is almost always interrupted by electricity pylons, roads and houses, whereas in New Mexico, Arizona or Utah, your gaze can wander endlessly over forests and mountains, untainted by signs of human civilisation.

However, it is only your eye that can wander so freely. Because in most cases, you're barred from stepping off the public highways, sometimes quite literally. For example, during a round trip through the south-west of the USA, barbed wire fences lining hundreds of miles of road very quickly suffocated that sense of freedom. Often what was fenced off was only sand and rock – as if someone would take something away! Privately owned land (and there is a great deal of it) is not open to the public, and there are frequent signs reminding you of that fact.

It was only when I was back in Germany that I came to appreciate the possibilities our woodlands offer to the public. Not only are all the paths at your disposal, but the entire forested terrain is too. Whenever you want to stroll into the undergrowth – please, go ahead! Nobody can stop you, unless it's one of the very few districts that are exempt. Nature reserves, national parks and smaller protected 'Bannwald' forests usually include territory that is off limits, where you are not allowed off the designated trails. However, since such areas make up a tiny proportion of the total forested area of Germany and are always clearly signposted, you normally can't go far wrong. Other exceptions are freshly planted areas of reforestation, where the vulnerable young trees are fenced off. As tempting as it might be to hop over the fence and take a cross-country short cut, when it's a protected area, please go the long way around.

One last no-go area is anywhere where logging is in process. Wherever you hear the roar of chainsaws or

the rumble of a log harvester, there is a genuine danger
to life. Walkers can't predict where trees – up to forty
metres in length – are going to land, and the undergrowth often blocks your view of dangers up ahead.
Therefore, when logging, or thinning, is in process, the
area is blocked off at a radius of hundreds of metres
from the actual site, with warning signs on the forest
paths and striped tape physically marking out the
restricted area. However, the overwhelming majority of
the forest is free of such restrictions, so you can really
immerse yourself in nature. It should also be said that
this right to roam applies only to walkers and hikers.
Cyclists and horse riders should stick to the designated
paths, while all other means of transportation are generally banned from woodland territories.

So, what is the best way to get a taste of the wooded
wilds? For the casual woodland walker, I would recommend the denser deciduous forests. The forest floor is
mostly free of vegetation, and your way won't be blocked
by low branches. It's a very different picture in coniferous forests, especially when the trees are planted very
close together. The dead lower branches of neighbouring
spruces, pines and Douglas firs can weave together like
interlocked arms to block your passage. I sometimes
have to turn around and walk backwards to force myself
through the closely knit branches without getting
whipped in the face or getting a needle in the eye. A
stroll in a deciduous forest is a much calmer affair. If
you encounter a patch of grass between the trees, you're
best advised to go around it unless you're wearing sturdy

waterproof footwear: the morning dew and rain will soak through shoes in no time, and even the latest hi-tech fabrics won't keep the water out for long on such terrain.

Blackberries are a challenge the walker will often encounter. Not the berries themselves, of course, but the sprawling and prickly brambles. These mesh together to form barricades sometimes several metres high. If you have to cross a thicket of brambles, then you'll need to adopt the gait of a stork, treading the tendrils down as you gingerly step across. It looks odd, but then who is watching you? If you rush, or prefer not to affect such a teetering gait, you may quickly find yourself ensnared. As the tendrils close around your ankle like a lasso, you'll struggle to free yourself from this involuntary embrace, and often enough, another step will have you tumbling into the thorns – ouch!

Beware too the danger of falling when walking on a steep slope. Not because of the struggle to stand upright – no, the hazard lurks beneath drifts of leaves or snow, and dead branches where the bark has rotted away, lying on the sloped ground. If you step on a rotten branch, your foot is prone to slipping downhill, sending you hurtling down after it like you're on a fairground slide. It's happened to me often enough. By the time I consciously register what I've stepped on, it's usually too late. I topple sideways, wave my arms about clumsily, then crash down on my side. It's best to avoid steep slopes in wet weather because of the compounded risk of slipping. The best paths to follow along steep over-hangs are the ones the wild inhabitants use themselves.

Since woodland animals are faced with similar challenges to us, they prefer to use well-trodden, even trails. These may be narrow, often little wider than thirty centimetres, but it's enough for a safe passage. On long descents, such game trails will run parallel to each other at regular intervals, so you can descend by simply stepping down one or two terraces, and then carry on safely, following your wild guide's beaten track along the side of the slope.

When you reach the valley floor, there is often a route across a stream. You've managed to keep your boots dry so far, so why change that now? With this aim in mind, most woodland walkers will attempt to jump from one bank to the other. This ought to be relatively straightforward, after all, since brooks and streams are rarely wider than a metre. And it's true most people should be able to jump that far. But the fact is that you're not necessarily on dry ground to start with. When a stream has relatively flat banks, the water tends to seep into the soil, resulting in small areas of boggy marshland. So, what might seem a risk-free leap will often end with you plunging knee-deep into a swamp, drenching your boots and leaving you shivering on the long walk home.

How can this soggy fate be avoided? First, look for a place where the bank rises up more steeply. There's a good chance there'll be some stones under the surface. Also, staying close to the trees and stepping on the raised roots increases the chances of keeping your boots dry. Of course, it's a doddle if the brook is shallower than the height of your boots. If you can see stepping stones, even just below the surface – then have no fear and step boldly into the water. Over time, the mud will have washed off these stones and they're usually as firmly fixed to the river bottom as paving slabs – though they're likely to be a little slippery. In all my walks through the reserve I manage, I have never yet sunk into the stream bed; however, I've often come a cropper on a soft, boggy bank. The only slight hazard is misjudging the depth, but even then, all you risk is getting a bit wet, so at least you shouldn't get muddy.

Boggy and muddy ground is always an issue in bad weather. Of course, walking boots are designed for tough use, but who wants to carry excess clods of mud around on their soles and to have to scrub the leather clean later? Not to mention the risk of mud seeping in over the tops if you sink in too deeply. To reduce this risk, you need to reduce the pressure that your boots exert onto the ground by increasing the surface area of your soles. For this you might, for example, use some branches that are lying on the ground. When you step on a branch, your weight is spread over a larger area. But do make sure that the wood is not too rotten, otherwise it'll disintegrate under you, and you'll find yourself sinking.

It's not everywhere that you'll find a branch lying around, whereas tufts of grass are more common. Every little cushion of grass juts out of the mud like an island and you'll find they're surprisingly stable as stepping stones, allowing you to go island-hopping with dry feet to the other side. However, this is true only for genuine streams or brooks, less so for swamps or bogs. On boggy ground, tufts of grass are poised on spongy peat moss and will become more unstable the further you venture into such a terrain.

And what if you don't fancy going off the beaten track? Trekking through the scrub and undergrowth doesn't suit every occasion. Heading off the footpath isn't ideal if there are two of you and you like to chat as you stroll. Conversation tends to become rather monosyllabic, as the passable route narrows and forces you to walk in single file. In these circumstances, it's sensible to maintain a slight distance between walkers, because of branches springing back when you push past; and that puts even more strain on the conversation. Besides all this, what's wrong with sticking to the official paths? There's still plenty to discover on them, not least the tyre prints of heavy vehicles.

Now, you might find it immensely irritating when you walk through freshly thinned forests and you find the loveliest paths have been transformed into rivers of mud. Why should hikers have to wade through mud up to their ankles just because commercial forestry is ruthlessly harvesting timber? I can understand the perspectives of both parties – walkers and nature-lovers, as well as commercial foresters. After all, with very few exceptions, the paths were only laid for the sake of transporting felled logs by lorry to the nearest sawmill. For financial reasons, the priority is not making the trails passable for leisure use; they're intended for trucks, and generally they are drivable even when they're thick with mud. In the past, wood was felled only in winter and would have been transported during dry weather when it was frosty. But climate change has turned our winter into a rainy season with temperatures generally above freezing.

This means that in the forest I manage there are more and more situations where there are only losers. We often put a halt to the transportation of logs in the autumn, when the paths get muddy with the arrival of the gloomy damp season. We tend to hope for a frost that will freeze the tracks for a few days, but this rarely comes. The timber deteriorates in quality meanwhile, prone as it is to fungal attack, and the buyer rightly fears heavy financial losses. The timber needs to be dispatched by March at the latest – by which time some of it might have languished on site for six months – before becoming irreparably damaged. The tracks become a mud bath and have to be repaired afterwards at great expense.

I've often been told by visitors that they had been rudely prevented from going for a leisurely stroll in other forests, in most cases by a gentleman in green who wound down the window of his SUV to announce that a certain route was off limits. If you have any doubt about the authority of such individuals, just ask them to show you their official ID. Often they won't have one at all, because they're game wardens, in the employ-ment of the local game tenant. Their green 'Hunt Protection' logo might look official in the windscreen.

However, anyone can order one online and display it in their car, just as you can buy similar window stickers saying 'Agriculture', 'Forestry' etc.; it's an easy way to feign a legitimate reason to be driving on forest trails. In Germany, the only truly official stickers are those bearing the imprint 'Forest' or 'Forestry Administration', along with the coat of arms of the federal state or city. The drivers of vehicles displaying such a badge will be able to identify themselves accordingly and indeed it is their responsibility to. However, most state employees with the authority to stop walkers rarely do so, but instead discreetly keep out of their way.

Many hunters take a different approach. They find it rather annoying when they've been waiting all evening perched in their lookout waiting for a shot, when a late forest visitor comes strolling by with a dog, perhaps off the lead. It means their entire wait has been in vain, and it is understandable that they clamber down, disgruntled. But there's little they can do; they have no legal grounds to report you for 'breach of the peace' or any such claim. All the same, who's going to want to stand up to an angry, heavily armed stranger? If in doubt, in the event of a confrontation, simply note the car number plate and back away politely. If the verbal confrontation is aggressive and your interlocutor has a rifle hanging over his shoulder (or in his hands), you may have grounds to press charges for coercion.

2. The Woodland Detective

I ENJOY WALKING IN THE WOODS TWICE as much when it's snowing. Firstly, I love a real winter's day, with the crunch of gleaming white snow beneath my boots. And secondly, this is when so many secrets of forest life are revealed, when the animal residents and visitors clearly leave their mark in the snow. Not that there's any guarantee of snow, of course, or indeed of finding footprints in it. The first cold snap in a season is particularly good for spotting tracks, when the animals still haven't moved into winter mode and they're much more active than they are later during prolonged periods of frost. The best time to start your snowy discovery tour is early morning, before the trails are scattered by the wind or melted beyond recognition by the midday sun. Take a camera with you and capture what you find, so you can look the footprints up at home with the help of a field guide or a suitable website.

In the summer months, fine mud on the footpaths can also result in some excellent tracks, with paw and hoof prints left like a wax seal. You can also determine roughly how long ago the animal in question passed by. The important factor is how recently it rained heavily. Rainfall rinses the tracks away or at least erodes the sharp contours, making them less easy to discern. For example, if you spot distinct deer tracks and you know it rained the day before yesterday, then the animal must have passed that way in the last two days.

It gets particularly exciting when you discover wolf tracks – not that you're likely to in Britain or mainland Europe. The first time I spotted some was in dried mud on a Swedish forest trail. I was with my family in the region near the border with Norway, trekking by canoe. How do wolves and canoes fit together, I hear you ask. We were canoeing along a series of lakes, and we had to cross the land in between at so-called 'portages', where you unload the canoe, lift it out of the water and mount it onto a two-wheeled trolley. In this way, we transported the canoe and all our luggage along several miles of secluded forest tracks through the hills.

It was while taking a much-needed break in our challenging hike that we wearily glanced down and were amazed to spot our first real wolf tracks. There were no other hikers in this remote area, just Sweden's largest wolf population at the time. Honoured to have glimpsed such a rare sight, we nevertheless pushed our canoe to the next body of water with a renewed burst of energy.

Why did I mention the absence of other walkers? Well, they are often accompanied by dogs, which makes it trickier to analyse canine and lupine tracks. Dogs and wolves are very closely related, after all, and their paw prints are very similar. I could never say with absolute certainty whether a footprint was made by a wolf or a large dog. The context can help, of course, including most tellingly what's been in the news lately. Every evening, across the region, hunters are out on watch from their mounted lookouts, and every sighting of a wolf is reported immediately and published in the media the next day at the latest. Large paw prints in regions where there have been no confirmed sightings are more likely to belong to their domesticated relatives. Those found within regions where wolves are known to have been seen are worth considering more seriously. In contrast to the zigzag tracking pattern left by dogs, wolves produce very straight tracks with their paw prints falling into a single line, one behind the other. With wolves, the hind paw prints tend to land on the print left by the front paws – this is 'direct registering'. For further confirmation, check for other muddy patches nearby: if the prints are those of a dog, the prints of its owner may also be visible.

If you find faeces, the difference between a wolf and a dog can be discerned more clearly. Pets are usually fed tinned or other kinds of pet food, and their excrement tends to be of a uniform colour and texture. Wolf excrement, on the other hand, reveals visible clues of what the animal had for dinner. Chalky bones are mashed up with animal hair, often black, from digested wild boar. If in doubt, you could collect a sample of the faeces in a plastic bag and send it to the nearest wolf expert or ranger, who can pass it on to for confirmation in the lab.

The second biggest predator, the lynx, has unmistakable prints. If you see cat paw prints of this size (and you are in a region where lynxes are known to roam), there can be no doubt. If you're unsure whether they are feline or if they could belong to a large dog, check for the paw shape: dog and wolf paw prints are symmetrical with the line of symmetry falling between the middle toes, while lynx prints are asymmetrical. Claw imprints are very rarely seen with the big cats, while wolves and canines tend to press their claws (or rather, nails) into the mud. If you are a cat owner, your feline friend may even help you to identify when a lynx is roaming in the area. A colleague from the Palatinate Forest said that his cat refused to set foot out of the door if one of his larger relatives was in the vicinity. To him, that was a sure sign.

While lynx and wolf tracks represent the jackpot in the hunt for animal tracks, fox prints are a consolation prize. You can tell the difference between them and the paw prints of smaller dogs, because the fox has a similar gait to his distant cousin, the wolf. Foxes 'direct register' like wolves, leaving a similarly straight line of prints. The position of the toes means the print is longer and thinner than a dog's broader paw print.

Another sign of the presence of foxes is their burrows. They're likely to be some way off the woodland footpaths, but if you go foraging in the undergrowth for mushrooms, you may stumble across a fox's den. Usually there will be several entrances, or exits, dug into an embankment. You will know if they are still in use if you see fresh scratch marks and if there is no vegetation on the recently dug-out earth.

However, such a dwelling could also be a badger sett and indeed it is hard to tell the difference without paw prints. (If there are prints, then they're easy to distinguish: badger prints look like small bear prints with the claws pointing forward.) Badgers dig deeper than foxes and hence dump more soil outside the entrance to their setts. A ridge may be visible in the earth leading to the hole, as they always enter and leave along the same path. You might also find some wool or other soft padding, which the badger has dragged to its sett to make its home nice and cosy. Unlike foxes who defecate everywhere, badgers return to the same spot to do their business. They even bury their faeces, producing a distinctive smell. They also use scent signals to mark out their territory, so if the area around a burrow is particularly pungent, it might well be a badger sett. But to make things more complicated, different species of animals often live in the same burrow system at the same time: for example, badgers, foxes and martens. Even if you can't identify the inhabitants, it is still an exciting discovery, as such constructions can be in use for centuries and may therefore be as old as medieval half-timbered houses.

Footprints, faeces and dwellings are just a few of the possible indicators. There will be clear evidence, for example, of where a wild boar has wallowed in mud. After a refreshing mud bath (which may itself show the imprint of the recumbent animals), boar rub themselves against certain rough trees, using them as scratching posts to scrub off not only the dry mud, but also hair, which then gets stuck in the cracks in the bark. On the way to their scratching back massage, the soggy animals spray muddy droplets over the vegetation, leaving a Hansel and Gretel-style trail of where they've been.

Certain indicators reveal the presence of animals in a subtler way. In the spring, beechnuts germinate in the old beech groves. With their seed leaves, the tiny seedlings look like little butterflies cautiously unfolding their wings. Sometimes you can see a whole clump of beech seedlings sprouting from the ground. But how can that be? Why would they all land in the same place? Beechnuts are heavy and – whether blown by the wind or not – they tend to fall directly down beneath the mother tree. From a purely statistical point of view, the seedlings should be spaced evenly around the trunk. You might reasonably expect to find two or three in one place, but ten or more? No, it's not a coincidence, but is down to the hoarding tendencies of squirrels or, more often, mice. In autumn, they stock up their winter reserves, so they can snuggle up under the blanket of snow with these oily seeds ready and waiting when they need a snack. So, the appearance of a whole cluster of seedlings hints at tragic events over the course of the winter: clearly a hungry fox has passed this way and helped himself to the hard-working little mouse, whose reserves remained abandoned in the ground to germin-ate in the spring. We can feel sorry for the mouse, of course, or we can see it the other way around: the fox freed the beech embryos from their enemy and thus ensured their survival.

We can see the traces of woodpeckers in a similar way. When they build their nest hole in a tree trunk, they don't go for a rotten old tree. Who wants a rickety home, after all? No, they often choose entirely healthy specimens, and to avoid getting a headache from banging away at the hard wood, they work on their hole in stages. In the intervals between bouts of chiselling, which might be as long as several months, the construction site is colonised by fungus, decomposing the wood and making it crumblier. Meanwhile, woodpeckers have other needs. They like to sip the sugary tree sap that rises through the trunk in spring. They are especially fond of young oaks, and to reach the sap they drill small holes in rows about ten inches long. The damage isn't too deep, but it does leave scars on the bark for decades that are almost ornamental in appearance.

The woodpeckers' search for insects is a little less painful for the tree. Insects only attack trees when they're dead or at least so seriously ill that the end is nigh. In the summer, when bark beetle populations are at their peak, the presence of woodpeckers clearly high-lights which trees are infested, as the beetle larvae are a delicacy woodpeckers cannot resist. Wherever there are juicy white grubs squirming under the bark, wood-peckers will be pecking and poking around until they have devoured the majority. This feasting frenzy results in the bark coming away over a large area, and the light-coloured wood that is revealed is a clear signal even from afar of a beetle infestation.

Even dead logs, slowly decaying on the shady ground, are an attractive feeding ground for wood-peckers, with over a thousand insect species laying their eggs in the rotting wood. The pale larvae often spend years munching their way through the crumbling wood fibres, before they pupate and finally emerge in the outside world as beetles, an adult life that spans a mere few weeks. These woodpecker larders are particularly worth investigating in winter, when there are no ants marching about and the flying insects are dormant over winter, hiding under the peeling bark. In their hunger, the woodpeckers hack and peel away at the dead wood, digging down to reach the protein-rich larvae deep inside: ample compensation for their hard work. When there's a particularly good haul to be found, you'll later find shredded strips of deadwood littering the ground.

The next category of animal signs might be better described as remains, and a small incident at our house – the forester's lodge – recently reminded me that they also belong in this book. I was sitting eating my lunch one day, biting into my cheese sandwich, when I was surprised to see that it was snowing. The snowflakes drifted gently to the ground – too softly, somehow. Looking closer, I saw they were in fact downy feathers. I got up and went to the window. The source of this precipitation of feathers soon became clear: it was a jay gleefully plucking a great tit in preparation for a feast.

Small tragedies of this kind happen all the time under the canopy of the trees, with birds falling victim to one of their many predators. Mammals that hunt birds include squirrels, martens and foxes, to name just a few. Among birds, predators include corvids such as magpies, jays, crows and ravens, as well as owl species such as the tawny owl or the eagle owl, and birds of prey such as the hawk or sparrow hawk. You might find a heap of plucked feathers lying on a tree stump – the tell-tale remnants of someone's dinner. Certain species also seem to like using tables for their butchering. It may not be possible to tell precisely which animal was the culprit, but dents and marks in the wood might be a clue that it was a bird rather than a mammal. Birds have no teeth, so while foxes, for example, can take bites as they eat, birds of prey have to rip their dinner apart with their beak, leaving notches or scratches in the wood where the beak pecked at the prey.

While you're keeping your eyes peeled for tracks, there's one more animal group you could investigate. What about applying your detective skills not just to animal tracks, but to human ones too? After all, these are the most common tracks you'll find during a forest walk. It can be fun to play detective. Look at the puddles, for example; a great way to see when the last vehicle passed this way. If the water is still cloudy, something has moved through it the same day, often less than an hour ago. A single tyre track suggests it was a 4x4, double tracks point to a timber transporter. If you see tracks further apart than a regular truck, it may have been a harvester, in the forest to fell trees and take the timber away. It can be as intriguing to investigate trails left behind by human users of the woods as tracking the footprints of the wild population.

3. Spotting Wild Animals

ADMITTEDLY, IF ALL WE SEE IS TREES, any hike in the woods can get a little monotonous sooner or later. Even the keenest forensic analyst can lose interest eventually in spotting the tracks of earlier visitors. What can really spice up a walk is spotting wildlife in the flesh. As a general rule, the bigger the animal, the rarer the sighting. There are two reasons for this. Larger animals need a more expansive habitat. A lynx has a range of over fifty square kilometres, while a wildcat can get by with five to ten. A fox roams over an area of less than a square kilometre, while a deer needs a much more modest 0.02 square kilometres. You see, a carnivore requires a much larger area than a herbivore.

The same applies to small creatures, of course, on a corresponding scale. In pristine forest, certain carnivorous spiders, for example, might compete with up to a hundred others within a single square metre.[1] If you lie down on the soft duvet of autumn leaves on the forest floor, you'll quickly find you're surrounded by these little predators. You'll see spiders out to catch flies, woodlice, springtails and mites, all of which are found in much larger populations beneath the blanket of leaves. If you want to watch these mini beasts closely, you should crouch down and concentrate on one square metre.

A small sand sifter (e.g. from a toy shop) and a magnifying glass will serve you well on an expedition into the microcosm. And you might want to bring a picnic blanket to lie on; the more comfortable you are, the longer you can spend observing and the more you'll see.

However, it can quickly get dull if you can't identify the species that are buzzing and crawling around you. Since there are so many, I recommend that you concentrate on one class of creature at a time and take along a suitable field book. Sticking with arachnids, for example, there are over a thousand species of spiders in Germany alone. That should be enough for some stimulating observations, especially as you may also find newcomers which are arriving in increasing numbers via global trade. Recently my daughter raised some loose tiles on her balcony and, to her horror, uncovered several venomous European black widows, usually found in more southerly climes.

With larger animals, there's the advantage that you can sometimes see them while walking. But since they are also the focus of human hunters, they are often extremely shy. Twice a year, however, there are occasions when deer are tamer. One is the mating season, when hormones obscure the senses and males in particular drop their guard. Rutting season can be a tourist magnet even in areas with heavy deer stalking, and in extreme cases, such as the countryside near where I live, you can sit on a folding chair at the side of the road and for two weeks a year enjoy a lively show of testosterone-fuelled stags roaring and rounding up their harem. The other time when you're more likely to see deer is the end of the hunting season. When the rifles go quiet at the end of January in Germany, word quickly gets around among the wildlife. The longer it is since the last shot rang out, the more fearless these hunted species become. The best sightings are shortly before the first of May, when the hunting season starts again (in many parts of Germany at least). Deer graze peacefully on meadows and at the edge of forests in early spring, barely aware of the presence of walkers, as long as they keep a distance of about a hundred metres.

I have already introduced the wildcat, but I should mention it again in the context of animal observations. Sightings of this species are a rarity, and there are three reasons for this. These shy animals stick to very sparsely populated areas where humans rarely tread, which means that you hardly ever come face to face with one in a region oft frequented by walkers. The second reason is their small number. Only a few thousand cats are spread out over tens of thousands of square kilometres (with many forest areas where none live at all). In those regions where they are found, they share their territory around the outskirts of rural villages with millions of domestic cats, and the tabbies among them can be similar enough to their wild cousins that it's easy to confuse them. Few of us can be certain it's a wildcat we've seen, but there are a few clues we can rely on.

First, there's the coat. In wildcats, the stripes are very faint and the colouring is a mixture of grey and ochre. The hair is longer and as a result the tail is bushier, with black rings and a black tip. The nose is flesh-coloured, and in terms of both height and weight they are slightly bigger than domestic cats. But as the stripes are more distinct in young wildcats (and of course they're smaller), it can be difficult to tell the difference between them and domesticated tabbies. The only way to tell for sure is genetic analysis. In winter, at least, there is one other clue: if you see a cat over two kilometres from any houses, it is much more likely to be a wildcat. A pet wouldn't stray so far in the cold; they'd stay close to the warm hearth on a winter's day. Wildcats have no choice other than to find a hollow tree to curl up inside.

A bird feeder is a simple and effective way to attract birds to your garden and increase the opportunities to observe wildlife. In the past, I was vehemently opposed to doing this in our garden as, after all, providing an artificial food supply distorts the species composition. Winter is usually the time when only the fittest survive, and the species that don't migrate have to make do with the scant food that's to be found amidst the snow and ice. If we help them through the winter with fat balls and sunflower seeds, a greater population survives than would naturally. In the spring, these dominate the breeding grounds and compete for insects with the migratory birds returning from the south. The returning migrants don't stand a chance. Nobody helped them through the winter, and just when they need to recover from their exhausting flight and start a new family, all they hear is the chattering of birds who have already staked a claim to that territory.

I was always against such an intervention – and I stress, I *was*. Because at some point I gave in to pressure from my children and against my better judgement I built a bird feeder outside the kitchen window. From that day on we found ourselves watching our visitors, captivated. And I soon found that even for me, it was well worth it. No sooner had I installed our bird magnet – a little red house built in the Swedish style – than we had a rare visit from a middle spotted woodpecker. This is the smaller relative of the great spotted woodpecker and to survive it needs very mature beech trees, over 200 years old. Fortunately, we have trees of this vintage in our reserve, where they are strictly protected. The reason the middle spotted woodpecker needs such old trees is very simple: the bark of young beech trees is too smooth for the birds to grip. It's only at an advanced age that the trunks develop the kind of ridges (or, let's say, wrinkles) that a woodpecker can dig its talons into. Although our property joins directly onto this old beech reserve, I had until that point never seen one of these rare woodpeckers. What a joy it was to now be getting regular visits from these delightful birds.

To stay with the question of feeding: why should we stop at birds? If feeding red deer or roe deer is problematic because it increases their numbers, why is it not a problem with birds? It's a real dilemma, which is gladly exploited by lobbyists. A few years ago, for instance, our local paper was full of reports of scores of deer going hungry and even starving to death. The plight of the animals was so great that they were even breaking into cattle sheds and stealing the hay from the cows. A colleague sent me a photo he'd taken on his mobile of a young deer plundering a garden bird feeder. In my entire career, I'd never seen anything like it. Should we not intervene? Was it time to take provisions of hay, beets and oats into the forest to help these precious animals?

The instinct to help is human, but paradoxically their hunger is caused by human meddling. As I outline later in the chapter 'Good Day and Good Hunting', it is hunters leaving out food for the deer that distorts the process of natural selection in the winter. Left alone, when a population of animals exceeds the level that the ecosystem can sustain (i.e. how much food there is to go around), the weaker ones will usually perish in the winter. As cruel as it sounds, starvation is a completely natural phenomenon. It is only in this way that the vegetation remains in balance with the herbivore population. Feeding animals out of compassion only leads to the survival of a larger than usual population, with the result that the balance becomes increasingly precarious and problems escalate. Parasites such as tapeworm also become more widespread in a denser population, weakening the affected animals. For this reason, providing agricultural products to feed wild game is prohibited by the authorities in most regions of Germany. The trouble is that monitoring is so haphazard that the regulations are often ignored – and the problem remains.

In the low mountain ranges, such as the Eifel region where we live, winter often starts with a few weeks of snow. This quickly spells the end for the weakest animals, which die of exhaustion. In the particularly harsh winter mentioned above, the hunting lobby made an impassioned call for fodder to be brought into the forest for the deer. Whether it was politicians they were addressing or even schoolchildren, they pulled out all the stops to pressure the authorities into making concessions. The feeding ban was eventually relaxed a little, although too late to be of use: it took so long to jump through the bureaucratic hoops that the problematic deep snow had long since melted. While feeding deer like this doesn't necessarily mean you'll end up seeing more of them, squirrels and also larger mammals such as wolves will become tamer through being fed. And for many animals this can be perilous, especially for wolves. Anyone who encourages wild predators to become accustomed to humans will almost certainly contribute to their downfall when they are subsequently culled by the authorities.

But there are legal ways to increase your chances of seeing wild animals, and one is even rather idyllic: if you approach on horseback you're less likely to be seen as a threat. Deer hunters tend to encounter very little game, because these animals have a healthy fear of two-legged predators – as we saw in the chapter 'A Walk in the Wild'. Ordinary hikers are not seen so critically, but trekking on horseback will give you the best return in terms of animal observations. Horses are herbivores after all and won't be seen by deer as a threat. Just as in the Serengeti, animals don't worry too much about other species of the same type and will casually carry on about their business. Even if a person sits atop this peaceful being, the rider will be perceived as part of the horse. This, together with the raised sitting position, results in a significantly increased chance of observing wild animals. Now, horses are big, and, I confess, I used to be rather afraid of them. One kick from a horse hoof and even the strongest bone is shattered. But now I get on well with horses, especially since we have had our own. I was personally never drawn to learning to ride, but several years ago my wife fulfilled a long-standing dream of hers and bought a horse. First it was just the one, but then we bought a second – Bridgi, a young mare – as some company for the first. But Bridgi also wanted to be kept busy, and so I overcame my fear and learned to ride. When I took her out hacking I quickly realised how different the forest looks when your vantage point is elevated even by this modest amount.

In fact, you also increase your chances of observing animals by car. You may have already noticed this from driving: because deer don't perceive vehicles as a threat, they will happily graze on the lush green slopes of highways and byways, resulting in a lot of accidents and roadkill. Sitting in the car, your vantage point is lower than on a horse, but the pay-off is that you're not weather dependent. The whole thing only works because it is illegal to shoot game from a vehicle, so wild animals don't associate danger with our moving metal boxes. The only catch is you're not allowed to drive through the forest. But a leisurely evening drive along local

country roads may lead to more sightings than hours
of waiting at a viewing point in the woods.

So far, we have only dealt with species that are easy
to see with the naked eye. It is tempting to limit
ourselves to the larger species, because all too often it
is only these animals which are counted when the biodiversity of a region is measured – and not only by
laymen, but also by professionals. The tinier creatures
that evade our built-in optics are granted very little
attention. Unfortunately, this is often linked to an unfair
hierarchy by which various species are valued. Eagle
against ground beetle, lynx against springtail – it's clear
in each case which one would win our sympathy. And
yet it is well worth taking a look down at the ground
and through a magnifying glass.

There was an exciting discovery recently in the
ancient woodlands here at Hümmel: a rare instance of
a *Trachodes hispidus* weevil. The Latin name perhaps
doesn't evoke much compassion, but this funny little
thing is sweeter than he sounds. It's a beetle with a long
snout like an elephant's trunk and along its back it has
scales that stick out like a mohawk[2] – the reason why
my pet name for it is 'mohawk beetle'. It can't fly and
usually it has no need of a quick getaway. As a typical
woodland species, its habitat is ancient ecosystems that
remain the same over thousands of years. If it is
disturbed, it pulls in its legs and plays dead; there's not
much it can do, as it can't fly away. Its brown hue gives
it the perfect camouflage amidst the carpet of leaves on
the ground or on dead branches, making it very difficult

to spot for the untrained eye. I find species like this particularly interesting because their occurrence suggests the presence of deciduous trees that derive from primeval woodlands. Unlike most of Europe, which has been extensively cleared, ploughed and grazed throughout history, before our ancestors began re-foresting areas, here at Hümmel we have ancient, undisturbed ground where the little mohawks feel at home. It's time to see more professionally led beetle walks on offer to the public, instead of just guided bird-watching walks – the timid little fellows are long due their time in the spotlight.

When the fauna is hiding out of sight, don't be disappointed – there's always the flora. And just because plants can't run away, it doesn't mean there's nothing exciting to discover.

4. The Right to Roam and Pick Your Own

WHEN YOU STEP INTO THE WOODS IN GERMANY, your rights are not limited to the right to roam. If there is something edible to pick, then you may get stuck in! It's not everywhere in the world where you have that right. And of course it comes with some limitations. Imagine you have a small garden with a strawberry bed. Picture it: the day the berries are ripe, a family of complete strangers strolls in through the gate and proceeds to fill a wicker basket. Your strawberry plants are stripped bare, your dreams of homemade jam shattered. This is of course strictly prohibited, but such a prohibition is limited only to your garden. If you own forest, on the other hand, you will have to tolerate ramblers and hikers not only strolling through but also picking all kinds of fruit. There is just one small restriction: each walker may only collect enough for one meal. And for a little decoration on the dining table, as you're also allowed to pick a small bouquet of wild flowers.

Now, one could argue that a forest owner doesn't deliberately grow blackberries, wild strawberries and mushrooms; these all grow by themselves anywhere that suits them. This is true, but they are nevertheless seen as a product of the plot of land, and so strictly speaking they belong to the landowner. However, because wood-lands occupy a large part of the landscape in Germany and heavy-handed restrictions would affect the population's ability to relax in the great outdoors, this is where landowners' obligation towards the public comes into play, a principle enshrined in the German constitution as 'Sozialpflichtigkeit des Eigentums', literally the 'social responsibility of ownership'. This unwieldy term conveys the principle that the interests of the individual must not be at the expense of the common good. But some-where there is a limit, and in the case of mushrooms this limit is often grossly exceeded. I have often seen minibuses parking at the edge of the forest and offloading a good five to seven people who stroll off with plastic buckets in tow. But as they stroll they comb the ground so thoroughly for edible mushrooms that there's not a single morsel left. They empty their buckets into washing baskets and then head off again, collecting mushrooms the whole day long like this.

Not only is this unfair to other mushroom pickers, it's simply against the law. Firstly, because it's vastly exceeding the limit of one meal's worth, and secondly, these teams of pickers are doing it for financial gain. Knowing that, for example, cep mushrooms (porcini) can sell to restaurants for up to fifty euros a kilo, you can quickly calculate how much one of these minibus teams can make in a day. I'd estimate that in the autumn, most locally sourced mushrooms served in restaurants are picked illegally in this way. A trivial offence? To me it is far from trivial, because greedy profiteers jeopardise the principle of freedom for all. What's challenging is catching and punishing offenders, because what is a forester going to do when a group of outraged mushroom pickers deny any knowledge of wrong-doing? All they can do is request to see the vehicle registration number; the outcome of checks is unreliable and in the best cases all they can do is hand out a double-digit fine.

Incidentally, mushrooms are very special organisms, which belong to neither the plant nor the animal kingdom; the taxonomy of living things now grants them their own kingdom. Like animals, they can't produce their own food and thus rely on consuming foreign organic matter. Like insects, their cell walls consist partly of chitin, but they lack a central nervous system. Several species of fungus interact with trees as valued life partners. They help trees to find water and nutrients by wrapping themselves around the delicate tips of their roots and even growing within them. Their cotton-wool-like consistency means they considerably increase the tree roots' effective surface area, helping to draw more essential nutrients into the tree. Certain friendly fungi help block toxic substances, such as heavy metals, from their leafy allies and provide an effective barrier against other aggressive fungal species.

But that's not all; trees communicate through their roots and can warn each other about an insect attack or an impending drought. However, since their roots cannot reach every single nook and cranny, the fungal network ensures the onward delivery of such messages. Scientists therefore speak of a 'wood wide web', the internet of the forest. Fungi can be royally remunerated for their services to trees: up to a third of a tree's total energy production is passed on to its hidden helpers, mostly in the form of sugar. A third is about as much energy as is expended on the trunk, while the rest is used to grow the twigs, leaves and fruit. The fungus consumes this concentrated energy not only for day-to-day life, but also to produce its own fruit. The mushrooms we collect are in fact fruiting bodies, just like the apples of an apple tree. The fungus itself stretches deep below the soil with its gossamer white threads intermingling with the roots of many plants. A fungus can assume gigantic proportions. The largest recorded specimen is an *Armillaria ostoyae* in the Malheur National Forest, USA. Affectionately named the 'Humongous Fungus', this record-breaking organism has spread over nine square kilometres and weighs 600 metric tons. It is therefore the largest known living creature on Earth and is estimated to be several thousand years old.[3] This particular fungus is not so considerate towards trees, however, since it kills them for food.

Have you ever wondered why mushrooms mostly form their fruit in autumn? The answer lies with the trees. After all, the fruiting bodies of most edible woodland species are produced indirectly from tree sugar, and for this, sufficient quantities need to be obtained from the tree. In spring and summer, the tree requires most of its sugar for itself to produce leaves, new shoots and fruits. By late summer, however, most of the trees have already stored away enough energy for the winter and the coming spring. Increasing quantities can now be

passed down to its underground allies, which are eager
to get on and reproduce. And in the process to put forth
the delicious 'mushrooms', only for them to be harvested
by you or one of these professional minibus crews.

Few other fruits of the forest are so popular that
they attract the same profiteering zeal. In this category
of forest produce we'd find blackberries, raspberries,
blueberries and sloes, along with hazelnuts and bitter
wild apples, which we can at least cook up into a tasty
jelly. But here we're faced by another moral question: if
we pick these, aren't we stealing from the woodland
inhabitants? Wild boar, deer, birds, snails and insects all
rely on this fruit and nut supply and struggle to meet
their calorie requirements without it. We've already seen
the answer with respect to mushrooms; the yawning void
left by the professional gathering crews can be disastrous
for nature. But if you take just enough for one family
meal and leave behind the ones that have obviously been
nibbled at, then you'll be leaving plenty for the animals,
especially as you'll only find a fraction of what's out
there in any case. With berries, things are a bit different,
because most species were not originally found in our
forests in the first place. Blackberries, blueberries and
sloes all need much more light than is generally found
under the canopy of primeval forests. It's only where
trees have been cut down or thinned out that enough
sunlight can reach the ground for these plants to thrive.
Without doing so intentionally, foresters and woodland
landlords have turned these clearings into a kind of
cultivated landscape where fruit grows that no one has

sown. The food supply for the animals is therefore larger than it would have been naturally, so there should be ample quantities even if you help yourself.

In the past, the situation was more problematic because people harvested quite different things. In the period after the Second World War, for example, fat and oil were in short supply, so people collected beechnuts in the woods. These are rare enough most years even without human intervention, and the animals urgently need these calorie bombs for the winter. But during the war, the rural population were ruthless in their desperation. Unwilling to wait for the beechnuts to fall from the trees by themselves, they went around the forest bashing the trunks with mallets. The serious damage to the trees was seen as a necessary consequence. Gathering firewood, especially brushwood which isn't useful for much else, was also common until after the war and caused widespread harm to the forest. Twigs are mostly made up of bark, which means they are particularly nutritious for their size. Bereft of this groundcover of twigs, the forest was starved, leaving nothing for the smallest woodland inhabitants to eat. People also took loose twigs and leaves from the forest to use instead of straw in the horses' stables, and the woods became so depleted that the collection of brushwood was banned. The ban remains in principle until this day. I say 'in principle', because in fact the practice has crept back in through the back door and now happens on an industrial scale. The magic phrase is 'forest residue', which refers to everything except the usable tree trunks. After a tree is felled, the bough, branches and twigs are bundled up by machines and left on the verge beside the track to dry. Along comes the shredder a few months later and the shredded wood chips are blown onto a truck. The harvest is driven to the nearest biomass power plant, where it is transformed into green electricity. For the forest, it makes little difference whether it's local villagers gathering bundles of brushwood day by day for their wood burners or whether it's mechanised brushwood bundlers (as the mighty beasts are known) doing it all in fully automated fashion. Except that the brushwood collecting during the war probably inflicted less damage.

Another product of the forest comes into its own just before Christmas. Moss is eagerly sought by children in Germany to decorate their nativity scenes. Now there's no harm done if you take a modest amount for your yuletide decorations at home, even if you do inadvertently bring some funny little creatures in with it, tardigrades, for example. Less than a millimetre in size, these minuscule beasts are among the most resilient creatures on the planet. When they dry out (for instance, in your nativity display), they pull in their little legs, curl up and become indestructible. They can endure any temperature unharmed, from extreme cold to extreme heat. Once the temperature returns to a more comfortable range and the tardigrade is moistened by a droplet of water, its legs pop back out and it goes about its business as though nothing had ever happened.[4] It wouldn't even notice a short trip into space. So as long as you put the moss back outside after use, then everyone's happy.

I am much less happy, however, when I come across commercial pickers. They stuff whole stretches of moss into onion sacks and fill small transporters to sell it at Christmas markets at the expense of the forest. I am constantly on the look-out for overexploitation on my patch and do everything I can to prevent it, just as I stop people from taking plants and potting them to sell. Such money-making enterprises jeopardise the common right to free access and everyone's right to gather fruit, mushrooms and flowers for personal use. It's not only in Germany, but also in Austria, Switzerland and especially in Scandinavia that we often hear calls for this common right to be abolished. Therefore, I do not see the action of reckless profiteers as a trivial offence.

It is worth mentioning Scandinavia, which prides itself on the so-called 'everyman's right', or the right to roam. As well as the access you're granted in Germany, in Scandinavia you're even allowed to camp for the night wherever you like. The only exception is identifiably private land in the vicinity of a house, but that still leaves extensive forests and the shores of thousands of lakes at your disposal. Fires are also permitted, except in conditions of extreme drought. My family and I have often experienced just how generous this arrangement is, including most recently on a hiking trip to the Sarek National Park, often referred to as 'Europe's last wilderness'. In the mountains of Lapland, we have the right to roam anywhere *and* to spend the night wherever we like. But we will only be able to experience this feeling of freedom while tourists and locals respond to this trust with respect.

5. Freshly Washed and Bitten

WINTER HAS MANY ADVANTAGES. The forest is quiet like it is at no other time of year. Walkers are few and far between, the mushroom pickers have long put their baskets away in the attic, and from the end of January onwards there's no one out hunting either. Best of all, there are no mosquitoes or biting midges at this time of year. Or rather, they're still there, of course, but they're not active; they're sleeping peacefully until the following spring. Even when it warms up, it's not until May that the populations of the little bloodsuckers are large enough to be noticed. A couple of warm, humid weeks can cause an explosive acceleration in their development: an egg can transform within fourteen days into an adult insect capable of flying and biting. A small puddle is enough to serve as a nursery for the larvae.

Mosquitoes and midges love air humidity. So they feel particularly at home when the warm summer sun rises at dawn over dewy meadows. They're less keen on dry air and shimmering heat waves, but there's always somewhere damp for them to escape to in the forest; the humidity is much higher where there is constant shade. If you're out walking when it's been particularly rainy and you're after somewhere to stop for a rest, don't sit down in the depths of the forest. The ideal spot is at the edge of a clearing, beneath the first trees next to the open air; you'll benefit from the drier air while still being able to enjoy some shade. Even better is somewhere with a bit of breeze, because mosquitoes and midges can't stand the wind. So much harder to steer towards your intended victim when you're constantly being blown away as you try to land.

The time of day also plays a major role. The sunlight is weaker in the mornings and evenings, meaning that the humidity is higher, while conditions are least favourable for mosquitoes and midges at midday. That said, the time of day makes less of a difference in the woods, which are especially attractive to mosquitoes and midges after rainfalls. It's also best to avoid washing your hair shortly before you set out walking. Mosquitoes and biting midges go wild for the shampoo smell of freshly washed hair and you'll soon find your scalp teaming with admirers. If you'd prefer not to go out with unwashed hair, the only thing that will help is a hat (which is going to ruin your hairstyle, anyway).

And then there are the horseflies. Just to be annoying, they like things the other way around: they prefer to buzz about in the blazing midday sun. If you leave the dark forest and head out into the sunshine to escape the mosquitoes, you run the risk of bumping into a swarm of horseflies instead. This happened to me when I was on a hike once with my brother. We were walking through the Eifel National Park, it was a beautiful day and the idyllic footpath meandered along next to a stream, surrounded by meadows. Over a stretch of

a few kilometres, my brother was so badly attacked that
we had to call off the hike. And horseflies are particularly
nasty. OK, they can't be blamed for their innate behaviour, but you have to be an extreme animal lover to
tolerate these stubborn, persistent little devils that
touch down so softly and then take such an excruciating
bite. It might be like a choice between the plague and
cholera, but if I had to choose between mosquitoes and
horseflies, I'd choose mosquitoes and the deep forest
any day. Horseflies hate shady places, so they won't
follow you into the woods.

If you don't have any choice about which path you
take (perhaps you're out with a group and the organiser
hasn't considered such subtleties), there are other ways
you can reduce the risk of being bitten. Bite-proof
clothing is one option. Outdoor clothing manufacturers
sell a range of tops and trousers in fabrics which they
claim are guaranteed mosquito-proof. I have put them
to the test several times on my Lapland trips. The
trousers and top lived up to the claims. However, by the
end of a long day hiking in the Swedish highlands,
I counted around forty bites by enormous midges (much
bigger than ours in Germany, which are only about one
millimetre in length). It wasn't the trouser fabric that
let me down, but the fact that when you sit down, your
trouser legs ride up and reveal a weak point. The pesky
insects homed in immediately and bit me through my
black woolly socks.

Another possibility is to resort to chemical repellents, which you can spray not only on exposed skin,

but also onto your hair and any non-bite-proof clothes if you're worried. But beware: some repellents dissolve plastic, meaning that any fabrics woven from plastic fibres will be damaged. And that's not all. Substances such as DEET (diethyltoluamide) don't just get on mosquitoes' nerves, they also get on ours – quite literally. Because these chemicals diffuse easily through the skin, they can get into the blood stream and then reach the nerves. Tingling and numbness are the lesser evils; there is also a suspected risk of their causing brain damage. It may be safer to choose repellents made from purely plant-based ingredients such as cedarwood oil. You'll get used to the pungent smell, but then it seems so do the mosquitoes. The deterrent effect is limited and wears off after just a few hours, so you'll need to re-apply. So what's the best option? Well, if we stick to mosquito-resistant clothing then we can use an effective but safety-tested chemical in moderation, applying it only at the most vulnerable exposed areas (ankles and lower legs, hands, face and neck). If it's just a question of mosquitoes, and horseflies and ticks aren't also a problem, then you might get sufficient protection simply from your air stream, the breeze generated by your forward motion. It's only when you stop for a longer break that you should be ready for an attack and defend yourself accordingly.

If you're camping in the forest, there's another insect you'll need to keep an eye on: red forest ants. Listed as an endangered species in Germany, these mascots of nature conservation are actually relatively scarce in our naturally occurring woodlands. However, with the triumphant advance of spruce and pine, and the transformation of our forests into managed commercial plantations, these warrior beasts have been able to gain a strong foothold here. They're not unlike the forest police, ready to step in and clear up any carrion and pests. In fact, anything that can't move fast enough risks being ripped apart by their sharp pincers and hauled by hundreds of eager workers towards their big anthill. Deep down inside the hill live the queens, tirelessly laying eggs and being fed. The largest anthill I've found in my reserve, now abandoned, is five metres in diameter. The nest down below is even bigger, because they tend to stretch out underground to a similar extent again.

Forest ants have a sophisticated climate control system: when it's too cold inside the anthill, they go out to warm themselves up in the sun, then crawl back in and pass on their body heat to the others inside. In winter, they stay deep inside and are only seen if they're disturbed by woodpeckers or wild boars rooting around for some larvae, which take advantage of the warmth inside. The resulting holes are then patched up again in the spring when the ants drag in a fresh load of pine needles. Pine needles are fundamental to their ability to construct a hill; they couldn't survive in a deciduous forest. Have you ever seen an anthill made of leaves? The primeval forests that used to cover Germany, which consisted mainly of beech, would not have been able to sustain these state-building insects. Nevertheless, these masters of civilisation are deemed worthy of protected

status, and here we come back once more to the phrase
I used earlier: 'forest police'. Above all, foresters value
them for doing away with bark beetles. And indeed, in
years of heavy infestation, when whole spruce forests
are demolished by these hungry beetles, there are sometimes a few surviving pockets of green. When you
approach one of these green islands, you'll see what has
protected it: in the middle there will be an anthill whose
inhabitants have happily polished off any beetles that
tried to come near. But they don't just attack species
categorised as pests. They eat everything, even highly
protected caterpillars such as those of the purple hair-
streak butterfly. Nature doesn't distinguish between
pests or desirable creatures.

Ants follow long 'mini highways' into the
surrounding terrain. In order to make faster progress,
they clear away obstacles in their path; if you look
around, you'll find the infrastructure of a rubbish heap
in the vicinity of the hill. The bigger the anthill, the
further the industrious insects swarm into the
surrounding area. And the further away you should set
up camp. Forest ants are not dangerous, but they're very
unpleasant. Unlike other species, they have no sting, but
instead bite as a means of defence. To make it even more
effective, they also spray the area with formic acid.
I can remember more than one occasion when a little
critter has found its way onto my shoes and then up
my trouser leg, only to give me a hearty bite (by which
time I've been driving) in a rather tender area – ouch!
If you think you might be at risk, pull your socks up

over the bottoms of your trousers, and if you do stop in the vicinity of an anthill, running on the spot will help. As long as your feet are moving, if they do end up on your shoes, they'll usually jump off again.

It is fascinating to get up close to an ants' nest, to see the colony scurrying around in crowds around the many entrances or warming themselves up in the morning sun, to watch them dragging anything and everything into their nest. If you've never smelled formic acid, you'll be surprised at how pungent it is. If you want to know, just rest your hand for a moment gently on a spot of ground where there are a lot of ants. They will bend their abdomen forward between the legs and spray your skin. After two to three seconds, shake off any ants that are clinging on and hold your hand up to your nose to take a sniff. The smell is so strong it almost hurts.

You can dodge out of ants' way, but you can't escape mosquitoes, midges and horseflies so easily. How nice it is when a hard frost comes and wipes out all the pests! At least that's what I often hear people say, and I'd take objection. Quite apart from the fact that, no matter how unpleasant, I wouldn't give them such a dismissive title as pests, it's also not true. Insects aren't in the slightest bit troubled by hard winters. Most of the world's mosquitoes live in the Arctic; it's hardly likely to be colder in our climes. Incidentally, next time you see a TV report from the far north, watch out for the detail in the background. Just think how many mosquitoes and midges are swarming around the cameramen, flitting in and out of the picture. What really bothers them is the same kind of weather that we find wearing: a soggy winter when the temperatures barely dip below freezing and constant rain that drenches everything; that's the kind of weather that spreads illness among humans and animals alike. When we humans are catching winter colds, the chilly midges are also plagued by fungal and bacterial infections which kill many of them off. Incidentally, the same applies to larger animals, with a damp spring aggravating the situation even further: the new offspring struggle to keep warm and are especially vulnerable to disease in the wet cold.

Back to the parasites. As we've worked through this list of woodland 'pests' did you notice we haven't yet mentioned ticks? Well, they're next on the list, and as they present an increasing risk to walkers, I shall dedicate an entire chapter to them.

6. A Ticking Time-Bomb

UNLIKE THE DAYS OF LITTLE RED RIDING HOOD, forests are now generally harmless, at least in Central Europe. The large predators are mostly gone from our woods (we will come back to the wolf later), and there are no longer bands of thieves lurking amidst the trees, waiting to attack passing travellers. Poisonous beasts are also rare in our parts, so our woodland landscape is a lot like a peaceful front garden. Our instinctive alertness for danger has little to react to these days. It's hardly surprising that we experience this instinct in other ways now. Because large predators are in short supply, we concentrate our fears on small ones instead.

So when we go out for a walk in the country, at least in summer, there's one question we're quick to ask: 'Are there likely to be many ticks?' These minuscule beasts have become the ultimate villain hiding in the woods, tiny yet fearsome monsters lurking in the undergrowth to sneak up and attack us. I'll be upfront and honest: I can't totally dispel your fears as in this case they're justified. These small arachnids are genuinely dangerous, although, to be fair to them, they can't help it.

My first experience with these parasites was back when I was an apprentice with the Rhineland-Palatinate forest administration. I was assigned to a training reserve for my first year of practical forestry. I was given all kinds of tasks to complete, which above all meant one thing: spending most of the day out and about. On my first day, I arrived dressed in my favourite colour: blue jeans and a blue denim jacket. I felt well prepared for practical work in hardy denim. But straight away I got some funny looks: an aspiring forest ranger dressed in blue? They'd never seen anything like it. Embarrassed at standing out like a sore thumb, I drove to a country sports store the following Saturday and got myself some proper traditional hunting breeches, a shirt with imitation deer horn buttons and an army jacket – all in olive green. The funny looks vanished, yet my attire was completely inappropriate, as I was soon to discover. My mother furnished me with some long knee socks she'd knitted to go with my short hunting trousers, which chafed uncomfortably in the hot summer sun. Anyway, I felt like I looked the part at last, and I strolled merrily through the undergrowth in a clearing in my proud forester get-up. My cheerful mood was quickly dispelled when I got home and took a shower. I noticed small dots all over my legs – they were covered in ticks! I immediately sat down with some tweezers to pull them out, and after getting a couple out, I did a quick count to see how many more there were. When I got to fifty I gave up counting and just got on with it!

In hindsight, I know that I made two rookie errors. The first was the clothes: ticks prefer to sit in the undergrowth up to about knee height. And this part of my legs was covered only with woolly socks with large holes which they could crawl right through to get to my skin. The second mistake was my choice of route. Clearings with grass and small shrubs are ideal grazing territory for deer. As deer are one of the main hosts for ticks, the grass will be full of these little bloodsuckers waiting for their prey to brush past.

In conclusion, I'd say that the danger of a tick attack can be drastically reduced if you plan your route and your clothes appropriately. If you stick to the footpaths, you are unlikely to encounter a tick. They can't pounce on their victims from the trees, as if they did the slightest breeze would blow these tiny arachnids metres from their target. No, they rely on direct body contact, which is possible when an animal, or human, strolls through the undergrowth. But don't worry, they're not lurking on every single shrub and blade of grass – in some places they would have a long wait until food passed by. Ticks can go without food for up to a year, but they would clearly prefer not to, so they choose their hunting grounds, or 'questing' spots wisely. The best places for them to encounter host animals are wild game routes – narrow trails through the undergrowth where deer or wild boar regularly roam. Once they've had their fill of blood, the female ticks fall to the ground on these paths to lay their eggs, so this is also where the young are to be found. Ticks can detect when a large mammal is approaching through ground vibrations and by the smell. They climb to the top of the blade of grass and prepare for action, their front legs raised, ready to crawl across the moment the animal's fur is within reach. Then the tick nestles in through the hair to find a warm patch of thin skin, ready for lunch. But it won't necessarily tuck in straight away; it could be anything up to twenty-four hours before it starts sucking the host's blood.

After all, it takes the animal a while to find the best spot to latch on comfortably. While it's poking around in a leisurely manner, looking for a cosy picnic spot, you still have the chance to catch it and flick it back into the grass. To make it easier to spot them, it's best to wear light-coloured trousers so that ticks stand out like little black dots. Depending on their stage of life, they may be less than a millimetre in size, so you'll have to look carefully. In my experience, you can catch 99 per cent of ticks if you stop and check your legs as soon as you leave a grassy area.

If you've missed one, though, it won't stay on your leg for long, but will go for a little hike to find somewhere more comfortable. The ideal spot is somewhere dark and humid, such as in the folds of your skin. Once when I was camping, I was aware of some strange little scratching sounds as I lay in my sleeping bag. It wasn't coming from the forest outside my tent, but seemed to be inside my ear. The mystery was solved by the ear, nose and throat doctor: a tick had made itself at home on my eardrum and had started to suck away happily. Having it pulled out with tweezers was excruciating, but at least there was an immediate end to the unbearable scratching noise. The moral of this story is that if you find any evidence of ticks on your legs, even if you think you've got rid of them, don't go to bed until you've checked every single patch of skin on your body. Better safe than sorry.

So, what should you do if one of these little scoundrels does get his teeth stuck into your skin? In short, you should pull it out as fast as possible. You might hear advice telling you to twist to the left or the right; ignore this! Ticks don't have threads like screws, so if you twist you risk only breaking off the body and leaving the mouth part still attached. Simply pull it straight up and out – that's the quickest way. It is important not to squeeze the body, as this pushes the tick's body fluids into the wound, along with any pathogens it might be carrying. These pathogens also enter the host's body when the tick starts sucking, as it injects a little saliva into the skin to numb the area and to inhibit blood clotting. If the animal carries any bacteria or a virus, it will inject these at this point (about a third of ticks carry one infection or another). And it's these infections that really cause problems for us if we get bitten.

So what are these infections? One is the spiral-shaped Borrelia bacteria, the agent that causes Lyme disease. In many cases, the body is able to defend itself against an invasion of these bacteria, but often it cannot, and that's when they begin their sinister work. In an ideal case, the infection triggers a bull's eye-shaped rash: a growing red circle around a large red spot at the site of the tick bite. An ideal case? Yes, this response is ideal, because then at least you can be sure you have an acute infection. A trip to the doctor, a course of antibiotics and you should be fine. But if you don't get this textbook rash? Then you don't know whether you've been infected or not. Either the body fought off the bacteria without any help, or you weren't infected in the first place. If the tick had only been on the skin for a few hours, it might not have started sucking yet. Should you still go to the doctor? The only way to be sure is to do a blood test to determine whether the bacteria in question have infected your blood. But if you spend a lot of time outdoors, you can't send off for a blood test every time you get a tick bite. Personally, if I'm having a blood test for something else as part of a routine check-up, then I request they also carry out a screening test for Lyme disease while they're at it. If you're concerned that you've been exposed to it, it is advisable to have your blood checked in the autumn, when falling temperatures mark the end of the tick season.

If you come into frequent contact with ticks, your blood may show a similar picture to mine: there is a high level of antibodies, which would usually be a cause for concern. However, the presence of antibodies can act like an immunological scar in the blood from a previous infection from which the body has recovered; antibodies don't necessarily indicate a current infection. Since I had never had any other symptoms, my GP thought I may be a rare lucky person whose body can fight off Lyme disease by itself. But one day I had a severe headache that lasted more than a week. I suspected that I had come down with the infection and a blood test at a laboratory in Berlin confirmed the worst: my blood was teeming with the dreaded spiral-shaped bacteria. The only treatment that could help was an extremely long course of antibiotics. Fortunately, I tolerated the drugs very well and the treatment completely eradicated the disease. Since then, I have been much more vigilant about ticks, because sadly having the disease once doesn't give you immunity against repeat infections. Unfortunately, there is no vaccination either, because there are several different strains of the Borrelia bacteria that can make us ill. Incidentally, the symptoms are not limited to headaches. In the later stages, the disease can cause inflammation of the nerves, which can lead to severe joint pain and even facial paralysis. Once the bacteria has spread throughout the body, it can be very difficult to treat, so if you suspect that you have unusual symptoms after a tick bite, you should consult a doctor immediately.

In certain areas, tick saliva can contain an even more unpleasant pathogen, the TBE virus, against which there is now an effective vaccination. The full name is 'tick-borne encephalitis' which means inflammation of the brain. When people are infected, 70 to 90 per cent don't experience any symptoms, while others suffer from flu-like symptoms. Only a few cases in a thousand lead to death. Only a few per thousand? When we consider the risks we safeguard ourselves against in everyday life, even this rate seems too high. But thankfully the virus is not widespread across Europe, so it is not a risk you need to worry about everywhere you go on this continent. In Germany, at-risk regions include

Baden-Württemberg and Bavaria, and there is an
increased risk in Austria and Switzerland.[5] In Austria,
there has been a drastic reduction in the number of
reported cases, but this was caused by the successful
vaccination of over 82 per cent of the population, rather
than due to a decline in the prevalence of the virus.[6] In
the Alps, the only protection is to aim high: at altitudes
of over 1,000 metres, you're out of the danger area.
According to the Swiss Federal Office of Public Health,
no virus-carrying ticks have been found in these moun-
tainous regions.

So, should you get yourself vaccinated, just in case?
There are strong feelings on both sides of the argument,
as always when it comes to this topic, and the risk of a
vaccine's side effects have to be weighed against the risk
of infection. The risk of a single bite triggering the
disease is very low. According to Germany's Robert
Koch Institute, even in the main regions where the virus
is prevalent, it's still only 0.1 to 3.4 per cent of ticks
that are actually carriers. But if you're like me and you're
regularly out and about in a high-risk region, then you
are likely to benefit from the vaccination.

We are seeing a long-term upward trend in the
incidence of diseases caused by tick bites. Why is this?
This is a matter of debate between scientists and the
game-shooting community. What has game shooting
got to do with it? There has been a dramatic increase
in wild game stocks in recent decades, because of
hunting, or more precisely because of feeding by gamekeepers. If you've ever seen a deer up close in the summer,

you'll know how many ticks latch on to an animal for their final supper of blood before they drop down to the ground, filled to bursting like round little peas, to lay thousands of eggs before they die. When they hatch, the young suck the blood of mice, hedgehogs, foxes and other mammals, and in the process become infected with pathogens. Populations of these smaller host animals do fluctuate, but to my knowledge there's been no long-term steady increase. With deer and wild boar, it's a very different picture: there's been a population explosion. The more deer there are, the more ticks – it's easy to see the correlation. But the matter is more complicated than it seems. The pro-hunt lobby claims that ruminants such as deer destroy the bacteria within their blood, so ticks can't be infected by them.[7] Also, when it's a female tick's final meal before laying her eggs, it makes no difference if it is infected at this stage, as the Borrelia bacteria are not transferred to the eggs. But with population levels swollen up to fifty times the natural level, the little bloodsuckers have much better chances of reproducing, and it is their thousands of offspring that become infected from their first meal when they feed from small mammals such as mice. This is how we have seen a starkly increased prevalence of Lyme disease (Lyme borreliosis) and TBE in the densely populated Central European landscape.

Let's remain with this question of the vastly inflated wild game populations. It is even referred to internationally as 'the German problem', because we now have one of the highest population densities of mammals in the world. It's true that you might not see much in the way of deer and wild boar, because they hide in the woods during the day. Animals hiding in fear, diseases spreading into the human population – what's going on in the natural world around us?

7. Good Day and Good Hunting

I SOMETIMES THINK OUR FOREST LANDSCAPE is a little like the former Iron Curtain. Every few hundred metres or so, you come across a watchtower manned by armed game hunters dressed in khaki camouflage. We've become so used to them in Germany that we barely notice them when we're out walking. For the wild inhabitants of the forest, it's a different matter, because those species that count as 'game' will know full well that those towers spell doom. A network of hundreds of thousands of high seats mean that the forest is covered from every conceivable angle, so there's barely a patch of forest left where an animal can't be fired at. Does that sound a little alarming? On the other hand, isn't hunting an ancient tradition that deserves to be preserved in its present form? After all, we have hunted since we have existed as humans. In days gone by, it was a matter of survival, of procuring meat, fur and bones, not to mention occasionally an act of self-defence against predators; it wasn't a question of morality for our distant ancestors. But what of wild game hunting today? In Germany, as elsewhere, it is a hotly debated topic, not least the question of whether it is still at all justifiable these days from an ecological point of view. In terms of social demographics, things look very different to the past; the number of hunting licence holders is on the rise and women are increasingly breaking into this traditionally male domain.

To place the topic in a wider context, let's consider the oceans. Hunting and killing whales are now generally met with international disapproval, even when it concerns species where stocks have recovered to something approaching a sustainable level. Thus, the hunting of sperm whales and minke whales prompts regular protests, and quite rightly, in my opinion, because we have long since ceased to rely on such animals for nutritious purposes. After all, global grain production yields on average five tonnes of yield per hectare per year.[8] With each kilogram of corn providing 3,500 kilocalories of energy and the average adult male requiring 2,500 kilocalories per day, a hectare-sized field feeds around twenty people per year. And a whale? About half the body weight consists of meat, which would be about five tonnes in a minke whale, i.e. about the same as a field of grain. But since the meat contains only 1,200 kilocalories per kilogram, theoretically only seven people could be sustained by a whale of this size for one year. One hectare of average farmland dispenses with the need to catch three minke whales. And because fewer than a thousand are killed worldwide each year, it should be relatively easy to meet the same calorie yield through other means.

There has been a lot of debate about whether indigenous people should continue whale hunting, with respect to their local customs. Personally, I would support the Inuit population of Greenland, for example, putting an end to whale hunting. In most cases, these populations live with access to the products of modern civilisation and the whale hunt is carried out with motor boats and firearms – methods that have little to do with ancestral traditions. Why have I digressed onto whale hunting? Because game hunters in Germany argue a similar case to that put by the Inuit or Greenlanders. Hunting is an ancient tradition and this is how our forefathers would have procured their food in times gone by. Objection, Your Honour! For the last few centuries, hunting has been the exclusive right of the nobility, a privilege not formally abolished in Germany until the revolutions of 1848. Since then the right has been linked to the land, and anyone who owns a small plot of forest or a field may also shoot deer on it. While this is still the case in Sweden, for example, in the German-speaking world a new law was supposed to replace this right just two years after 1848. According to the revised law, the right to hunt applied only to contiguous areas of land of over 0.75 square kilometres. But which smallholders or individual farmers owned that much land at the time? All smallholdings were forcibly merged into coopera- tives, which leased out hunting rights back then as they do now. And who could afford the expensive lease? Rich nobles, who returned through the back door as the only ones with the right to hunt, having momentarily lost this privilege in the revolutions. In principle, there has been very little change to this status quo. On average, a hunt costs up to 50,000 euros to manage per year,

once you factor in game wardens, battues and repara-
tions paid to farmers for damage caused by the wild
game. Only a very small number of hunting licence
holders can afford fees of that magnitude, so wild game
shooting is to this day regarded as an elite hobby.

This brings us back to the basic argument asserted
by the hunting lobby: hunting is an old tradition. That
may be true for a tiny fraction of the population, but it
has never been for the overwhelming majority. For thousands of years, our traditional method of feeding
ourselves has been farming, not shooting woodland
animals. Many of the customs of German huntsmanship,
with its ornate and glorifying language (blood, for
example, is called 'sweat'), have only been widespread
since the Third Reich, when the Reich Master of
the Hunt, Hermann Göring, decreed every hunter had
to follow complicated rites and horn-blowing rituals.

The trophy cult also flourished just eighty years
ago. Impressive stag antlers and long wild boar tusks
became the ultimate goal. In order to breed animals that
would provide such prized characteristics, certain
animals were spared so that they could pass on their
desired genes. A massive winter feeding programme
made sure that as many animals as possible survived
and that there was always a large stock and a good
selection of animals for the marksmen. The mounted
skull and antlers were then displayed at trophy shows
and evaluated according to a complicated scoring system.

And nowadays? Unfortunately, not much has
changed since the Nazi era. The laws and regulations

still reflect the old ideas about breeding, focussed mainly on winning something impressive with which to decorate your living-room wall. Each to his own, you might say; after all, it's probably not the strangest hobby out there. Meanwhile, the number of large herbivores has increased to levels that are, as already mentioned, up to fifty times higher than natural population densities. The impact of this can be seen in our forests, which are being ravished, especially the young saplings of native deciduous trees. The buds of cherries, oaks, beech or ash trees all get devoured, along with seeds such as beechnuts or acorns; everything disappears in gigantic quantities into the hungry stomachs of wild game animals. As a result, fewer and fewer deciduous trees are growing, and many forest owners can only keep the forest going by planting spruce and pine. Deer don't like to graze on them, because of the bitter taste from the resin and essential oils, not to mention the prickly needles. But with the help of these tree species, foresters can at least build up their tree stocks and create the illusion of a forest.

Hunting is also the reason that parts of the forest are fenced off. In many places, deciduous trees can only grow behind bars, and a colleague of mine even coined the term 'prison forestry'. Within the fenced-off reserves is a sort of harmonious idyll, not only for the trees. Rosebay willowherb can grow here in dense quantities, an impressive, tall wild flower with delicate, bright pink flowers. It's a particular delicacy for deer, which is why in Germany you'll rarely come across it outside the protected areas.

Incidentally, once the fenced-off areas are removed from the deer's potential grazing ground, the pressure becomes all the greater on the remaining territory. And because deer will always look hungrily into the fenced-off paradise of plenty, they will attempt to break in wherever they find the smallest opportunity, such as if a tree falls down in a storm and crushes the wire fence. Then it's often the case that not even hunting hounds can drive them out of the protected area, and the fence – now useless – has to be dismantled.

When it's not worth putting up a fence, because there are only a few beech or oak trees to protect, rangers tend to erect other monuments. What looks from afar like a military cemetery is actually a plantation of saplings each surrounded by a 'tree shelter' or 'tree wrap': a plastic tube that's open at the top and protects them like a miniature greenhouse. The disadvantage is not only the cost and the effort involved in wrapping each sapling; strong storms and snow can knock the casings over and, with them, the tree inside. Besides, as soon as the tips peep out of the container, they get nibbled all the same. Deer are very resourceful when it comes to getting hold of the best buds at the top. If they're out of reach, because the sapling is already two metres high, they'll simply knock the stem over and break off the top.

Other modes of defence against hungry deer include treatments applied directly to the buds. It's the unpleasant tastes that puts them off, not unlike treatments we use to stop us biting our fingernails. It could be chemicals, or it could be sheep's wool draped around the buds; either way, it means a great expense. Forest workers have to visit and treat every endangered tree once or twice a year – that is a lot to finance. In the wood I manage, before we changed our approach, it was costing up to 75,000 euros per year, which long-term would have been crippling for a community with barely 500 inhabitants.

And even if all these efforts are successful in helping more deciduous saplings to get established, there are nevertheless huge gaps in our woodlands. Because when a lack of funds means foresters can't go to these great lengths to protect the saplings, the ground becomes overgrown with shrubs and bushes or self-seeded coniferous trees. And yet the solution is simple: there should be a ban on feeding wild deer and there should be much stricter penalties for shooting large predators. This would bring down deer stocks to a level that would give young deciduous trees a chance.

Apart from the ecological and social impact of hunting, there is also the culinary dimension. This is something I got to know well at the start of my professional career, when I was an officer in the state forestry commission. Back then, one of my tasks was to sell on game meats shot in the state forest, including venison and wild boar. The rangers provided the animals to the forestry commission, where there was a cooling chamber in the basement, like a kind of oversized fridge. The animals were hung there to mature, with the fur and head still attached, only the entrails removed. Once a week, a game dealer came from Cologne came to take it all away. For me, it couldn't have been easier, as this crafty middleman simply bought the lot – and that's saying something. First of all, I feel I should fill you in on what happens to the animal after it's shot, before it lands on your plate – that is, if you eat game at all. And if you do, you might change your mind once you know a bit more about the process. Because what happens to the animals would be enough to have a butcher's shop closed down for health and safety if their sanitary standards were this low. First, I should clarify that the following information applies only to a fraction of the venison that we eat, but this fraction still exists, and you'll only know if you're unlucky when you take the first bite. Unfortunately, I can't spare you some rather unappetising details.

Let's start with the shot. If it's a good hit, the animal is killed instantly; this is a successful 'chest shot'. It's better for the animal, but bad for the meat. After all, in slaughterhouses, the animal is stunned and then left to bleed to death, in that the heart is still beating while it dies gradually. In the case of game hunting, the lethal shot hits the chest cavity in the region of the lungs. The animal is killed swiftly, but the blood stays in the animal. There's nothing harmful about it, but this is what is partly responsible for that distinctive gamey flavour. Unfortunately, it often happens that the shot is not so precise and instead pierces the stomach and intestines. Their contents are pushed by the impact of the bullet into the surrounding meat and once you've smelled that ripe aroma, you'll know why it is that some game meats taste so 'wild'. Even less appetising are the shards of lead bullet that are scattered through the flesh, like the pellets from a shotgun. You can of course buy lead-free ammunition, but some people are still working through their old stocks in their firearm lockers and it might be a while before they're all used up – it could take years.

And if lead isn't your preferred condiment, I doubt what follows is going to whet your appetite either. It's far from ideal if there's any delay before the slaughtered animal is found, especially in the summer months. If it has lain in the warm sun for a while, then the deer hunter will have to decide whether the meat is still usable. Sounds like something of a conflict of interest, right? The slaughter also takes place on the forest floor, with no water to wash anything down, and often in early twilight. When you think of quality foodstuffs, do you not picture the tiled, refrigerated rooms of a butcher's shop? Can you imagine if pigs and cows were being gutted right behind the supermarket in the car park? I can't picture their customers staying put. Of course, there are many game hunters and foresters who adhere to the highest hygiene standards and the products they supply are flawless. It goes without saying that there are regulations prohibiting the unhygienic processing of game meat. But even if the meat inspector checks the carcass over and gives it the all clear, all this prevents is the gravest health hazards. In my previous job, I frequently witnessed what can still slip through the net.

One game dealer I came across was particularly generous. He wiped off the mould that had clearly been forming for a few days, and was good enough to agree to buy, albeit with a slight discount, some particularly pungent pieces. 'Perfect for a venison terrine,' was his terse reply, when I asked him how he planned to sell meat in that condition. Pâté with a distinct 'gamey flavour'? Yes, I had tasted something like that before, and now I knew I never wanted to touch it again. Because it's not just remnants of the entrails or the meat festering in the summer heat that creates that intense flavour. Rutting animals are not excluded, but are also sent to market despite being pumped full of hormones. Stags that are high on testosterone taste as strong as boar, where the overpowering male notes penetrate so deeply into the fat that if you touch the meat, it's hard to get the smell off your hands even with soap.

Depending on the region and the species, you might even find a hint of radioactivity added to the unsavoury conditions. Even now, Chernobyl is still there to greet us, if you order freshly shot wild boar in Bavaria. It's a topic that the hunting community is obviously keen not to publicise and even the authorities are rather hesitant to release quantitative data.

Nevertheless, some unpleasant data is accessible, as demonstrated by a request for information from the Green Party members of the Bavarian state parliament. The state authorities' response was telling. Approximately 230 grams of radioactive caesium fell over West Germany, the former Federal Republic (before reunification). This tiny

amount irradiated forests to such an extent that the officially permissible limit (600 becquerels per kilogram) can
regularly be found to be massively exceeded, especially in
the meat of wild boar.[9] It's helpful to compare this with
the average values for other foodstuffs showing light radioactivity (as all organic substances do), which are so low
as to be negligible at 0.1 becquerels per kilogram.[10]
According to information from the authorities, the radia-
tion level detected in some animals is, at more than 10,000
becquerels per kilogram, up to fifteen times the permissible
level, and these have to be disposed of. Why is it the case
with wild boar? They feed differently to deer, eating a lot
of mushrooms and truffles, for example – fungi which,
depending on the species, collect the metals in question
and concentrate them in their tissue.

What's controversial about the data is that measurements are taken only on random samples, meaning
that many animals are consumed that have not been
examined. Why aren't they? I imagine it has a lot to
do with the concern that the market for wild boar meat,
which is not a given, would then finally collapse.

Let's suppose, however, that game meat was always
processed in the most hygienic way possible. In that case,
is it not justifiable to obtain meat from animals whose
stocks are not endangered? After all, they have spent
their lives in freedom and not in the narrow, excrement-
smeared stalls of factory farming. This way, we can eat
meat while also reducing animal suffering. And if we
have shone a spotlight on the negative aspects of hygiene,
lead contamination and radioactivity, then we should also

be fair and mention the other forms of contamination which wild game is less exposed to than farmed meat. Medicines such as antibiotics or deworming agents (which act like insecticides) are not found at all in wild boar and venison. And as with other products of the forest, we should ask ourselves whether the benefits of consumption don't outweigh the negatives. Whether it's wood, berries, mushrooms, fish or venison that we're gathering or hunting: as long as we do no harm to nature, should it not be seen as a legitimate and ecologically justifiable form of food and fuel production?

As long as we do no harm. That is the crux of the matter. But a harmful impact can also be caused by genetic modifications, and I don't mean genetic engineering. When we humans place other living creatures in our service, we exert a sort of evolutionary pressure. Animals we hunt begin to adapt to survive. Adaptation means that they are no longer easily captured, and what could hinder us more than camouflage that makes us struggle to see them? Of course, wild deer can't don invisibility cloaks, but they are still very good at staying out of sight. They've made quite an easy transition, choosing not to come out in the meadows and fields during the day, but to hide in the bushes and the deep forest. Often you hear the claim that deer are nocturnal, but this is by no means true. They simply relocate their day-to-day business to where they are out of view, because as typical herbivores they depend on food intake around the clock, interrupted only by breaks to lie down and ruminate. Hiding out of view would make little sense with relation to their earlier predators, wolves, because they are armed with sensitive noses and ears to detect their prey. No, game has adapted over the centuries to the human predator, and the paradox is that on the one hand we have incredibly high densities of game species, while on the other hand we hardly see them at all.

Even with trees (which in a sense are also our prey), this natural selection can be seen in the timber harvest. There are many features we don't like but which trees need in nature. Spiral growth, for example, makes some trunks look like wrung-out towels. This twisting has the same purpose as springs in a suspension system: it lets the tree sway back and forth in a storm without snapping. Unfortunately, the boards sawn from such wood also twist when dried, rendering them useless. For this reason, foresters tend to have trees with this characteristic felled while they're still young and sold as firewood – a little twisting makes no difference when it's being burned. The only trees that make it to become thick and old are the flawless straight ones, which promise a straight grain and therefore the highest profit. So, these model students are also the ones that reproduce and pass on their genes to the next generation. Other similarly undesirable characteristics include forked trunks (where the trunk divides into two boughs) or bends, which make it difficult to saw straight timbers.

Thus, the whole forest gradually changes according to our needs, but the trees are imperceptibly transformed into genetic cripples. The genetic material that is responsible for spiral growth, etc., is often linked to other properties that are bred out at the same time. There is still much research to be done into what genetic properties might be lost along the way, so the future resilience of forests is like a game of roulette. Trees are in the spotlight for profit-making forest owners, so they at least recognise the genetic Catch 22. But what about mushrooms? The ones that end up in the frying pan are fruiting bodies which are robbed of the chance to reproduce, so the ones that get the most opportunity to multiply are the specimens or species that we don't want to eat.

What should we draw from this? Would we be better to keep our hands off? It is clear we humans have the capacity to exert serious changes in the ecosystem. And the gravest impact on the forest is that left by modern forestry. Where spruce plantations stand instead of the former beech forests, thousands of native species have been displaced. When wild animals are artificially fed for the sake of maintaining hunting stocks and are selected for shooting on the basis of the size of their antlers, then the entire bioceonosis, the ecological community, also changes. The swollen wild boar stocks, for example, have led to the local extinction of certain rare species of freshwater snails. Boar like to wallow in the mud, and even in dry summers in Germany, it's easy to find them in woods where springs seep from the ground. If the sprawling boar populations turn almost every little damp patch into a spa for the larger mammals, then the tiny snails, which need clear cool water, don't stand a chance. The millions of hungry deer, which devour the tasty rosebay willowherb in many places, are having a similar impact. I believe we need to do more to protect nature, above all against such pronounced changes. Mushroom pickers have much less impact on species composition than, for example, timber harvester drivers who crush the soil with their enormous machinery, smothering most of the life down below.

Appreciate small pleasures, refrain from great sins – that's my motto. Anyone who enjoys some freshly picked mushrooms with their dinner or homemade raspberry jam also has an interest in preserving the ecosystem. Even firewood can be environmentally sustainable if the wood comes from certified organic businesses that promote native deciduous forests. So that all the wonders of the forest, of which to this day we have only a rudimentary understanding, can remain undisturbed, we need to set aside a certain proportion of land as protected areas. So far, not even 2 per cent of the forest is left untouched and protected from commercial use; for a rich industrialised nation this is a woefully small area.

But to return again to the question of gathering the products of the forest. Here lurks another danger we haven't yet touched on: fox tapeworm. Does the risk of contracting this nasty disease perhaps not make everything we've said obsolete, when we see yet another reason to advise against taking food from nature?

8. The Fox: A Friendly Foe

FOXES ARE NO LONGER MUCH OF A THREAT. Well, they take the odd chicken here and there, as has happened to us a few times at the forester's lodge. One morning in March, I looked out of the window and said sleepily to my wife, 'Look, it has snowed again!' The lawn was white, but my wife brought me back to reality. 'That's not snow, it's feathers.' It turned out our jury-rigged chicken run was not really fox-proof; one had managed to grab the entire flock that night and dismember them beneath our bedroom window. An event that used to be a severe blow for the rural population is nowadays just a nuisance. And the biggest threat associated with foxes, which used to be a source of widespread fear and dread, no longer exists in Central and Western Europe: rabies.

It's a disease that merits the fear it used to inspire. If you're infected with the virus, usually from the bite of an infected animal, you won't be aware of anything at first besides the bite wound. It's only a few weeks later that the flu-like symptoms emerge and by then it is too late for treatment. The virus has settled in the brain and in the neural pathways by this point, where it develops its fatal power. Among other things, it affects the behaviour in infected humans and animals, causing outbursts of anger and biting. Death comes after a few days of suffering. Even if relatively few people were infected, it was clear that this terrible disease had to be eradicated. Enormous efforts were therefore made in many countries, including Germany. For decades, these were focussed on simply trying to eradicate the main transmitter of the virus, the fox. Every means was deemed justifiable with regard to the battle against rabies, and animal welfare was thrown out the window. Foxes were mercilessly shot and the animal was considered a legitimate target at any time of year. It was accepted that vixens with cubs would also be killed. The helpless cubs were supposed to starve to death in their den, and if the mother wasn't found, the offspring would be dug out and slain. It was soon found, however, that the fox population could not be controlled in this way as the animals reproduced all the more fervently in response to this brutal culling operation. Even when their dens were smoked out with toxic gas, nothing seemed to wipe out these resilient creatures.

The authorities finally had a change of heart and invested instead in a mass vaccination programme. But how do you catch wild animals to immunise them when being viciously hunted has taught them to be extremely wary of humans? The solution adopted in Germany was to use bait with capsules containing the active ingredient. To quickly eradicate rabies from very large territories, these little packages were scattered from planes. One landed on our bin at the forester's lodge so that I had the opportunity to examine it up close. It consisted of frozen fish waste, encased in a small plastic container.

Inside was the oral vaccination serum, which the fox would swallow along with the fish. I wondered back then what the greatest risk was: rabid foxes causing a small number of infections in humans per year, or blocks of frozen fishmeal falling on hikers (and forest rangers) from the sky. Thankfully, I never did hear of anyone being hit in the head by falling fox bait.

Since pets such as dogs, cats and horses were also vaccinated against the disease, the Europe-wide programme succeeded in completely eradicating rabies. Or, at least, almost completely. Because other mammals can transmit rabies, not only pets and foxes. Even with quarantine in place, the risk remains with dogs imported from countries where rabies is still present. The danger might also come from the air, as bats can transmit a closely related virus and – in extremely rare cases – they have been known to bite people.

At least the fox is now out of the firing line, but it's not out of the conversation altogether when we're talking about diseases. Rabies was hard on the poor old foxes, but it was the main regulator of the population. Without this restraining factor, the fox population swells, bringing them into closer contact with each other, which in turn increases the transmission of other diseases. A wave of illness can then lead to a population collapse. When most of the local population dies out, the few remaining animals end up being relatively isolated and the greater distance between them means they avoid getting infected. When the infection is stopped in its tracks, the population can recover again for a few years, before the cycle starts again. Rabies is now history in Europe, and the fox populations have made a strong recovery everywhere. Particularly strong, because most of our foxes are now in rude health.

But nature being the way it is, there's always something new around the corner to keep things in check. As soon as one bug's out of the picture, another comes along to take its place. And the gruesome epidemic that's come to take the place of rabies is the small fox tapeworm. This parasite immigrated from Siberia with infected muskrats and has spread widely in Europe since the 1990s (although, at the time of writing, it has yet to reach the British Isles). In the state where I live and work, Rhineland-Palatinate, the local Agency for Consumer and Health Protection reports that more than 20 per cent of foxes are now infected.[11] And our region is by no means the worst hit by the epidemic.

Have we done ourselves a disservice by swapping rabies for fox tapeworm? Let's take a closer look at what these tiny parasites get up to. Their eggs are certainly minuscule – as tiny as dust particles. They are excreted from the adult worms, which are also not very big – barely three millimetres in length. So it's hardly surprising that an infested fox might carry up to 200,000 worms in its intestine at one time.[12] This results in an awful lot of eggs ending up in its faeces, which is dotted around the open countryside. Mice have a little nibble on the droppings and unintentionally pick up the eggs. The larvae hatch within the mouse's body and infect internal organs such as the liver. The critically ill rodents become too weak to scarper when a fox pursues them. And thus the parasites reach their destination, the next fox's stomach. As the mouse is digested, the tapeworms emerge from its body like attackers from a Trojan horse. Their end goal is the fox's intestine, where they make themselves at home and live off their host's partially digested dinner. The parasitic tapeworm doesn't really harm the fox, which continues as an unknowing host for years, feeding the worms and un-wittingly helping them to multiply.

The whole thing becomes dangerous when humans swallow the minuscule eggs instead of the mouse. The same thing happens to us as happens to mice: the larvae settle in our internal organs. However, many years can pass before the first symptoms make themselves felt – valuable time in which the condition could have been treated. You can't completely defeat the tiny larvae; they can only be controlled by lifelong medication. Left untreated, however, the infestation is almost always deadly. That's the ideal outcome for the parasites: the host should be rendered slow and weak, so that the fox can catch him more easily. But as humans are not the fox's prey, in this case they are what scientists call a 'dead-end host', with the parasite having no way to move on to its 'definitive host', the fox.

The worst thing about this perfidious parasite is that, unlike rabies, you don't even notice you're infected. Rabies is transmitted via an animal bite, so if you're bitten and you know you're in an area where rabies is present, you should seek medical advice without delay.

Whereas, with tapeworms, you might remain unaware for years that you have unwittingly ingested some eggs. The minuscule specks are present mainly in fox droppings, but can also be found as deposits on berries or mushrooms. When a fox licks itself clean, specks of dusty faecal matter can scatter onto surrounding plants.

The official advice is therefore not to pick fruit from near the ground and to cook everything thoroughly just to be sure, because the eggs can withstand extremes of cold and heat up to 60°C. The danger genuinely exists, but the question is how serious the risk of infection is on your next walk in the woods. If you find some red, aromatic wild strawberries, should you only eat them if you cook them up into jam? What about plump, juicy blackberries? Should you never pop one straight into your mouth? What about stopping for a sandwich perched on a tree trunk? Should you eat anything without first washing your hands, in case you touched a flower or a leaf that might have had eggs attached to it? (And how are you supposed to wash your hands in the forest?) In future, you'll always hear an unmistakable voice in the back of your head reminding you, 'Fox tapeworm!'

For an objective analysis, it's helpful to turn to the data collected by the Robert Koch Institute, where they record and evaluate incidences of notifiable infectious diseases. The institute reported 112 human cases of small fox tapeworm in Germany in 2014. In eighty-three cases, there was information about the country of infection, that is to say the place where the eggs were thought

to have been ingested, for example, while on holiday. It
turns out that forty-eight of the infected individuals had
brought their unwelcome guests from abroad. Is
Germany therefore comparatively safe? The Institute's
researchers point out that the reported data reflects only
part of the reality, since according to their estimates,
two-thirds of the infections go unreported. So records
of around 100 reported cases might reflect an actual
figure of around 300.[13] Now, you could compare this
with the number of traffic fatalities (3,475 in 2015 in
Germany)[14] or with the number of people hit by lightning (about 50 per year),[15] but that isn't necessarily
reassuring. But there are other things you should
consider; your own body, for example. Since the fox
tapeworm is not aiming for us as its target host, it isn't
optimally adapted to our bodies. This means that our
immune system is able to fend off most attacks by itself.
The number of infections has risen in the last ten years
with the rising fox population, but seems to have levelled
off recently, according to the official statistics. Perhaps
it's because the fox population has also stabilised at a
steady level.

But perhaps we are focussing too much on the fox.
He has distant relatives who may pose an even greater
threat: domestic dogs. Among them, there are true
mouse lovers who rush to every hole in the meadow
and successfully dig up the unfortunate rodents. Clearly,
pet dogs can also get infected if they eat an infected
mouse. For our four-legged friends, it's as painless to
host tapeworms as it is for the fox; after all, the fox

tapeworm doesn't want to kill its definitive host, but to make use of their bodies indefinitely. Dogs, like foxes, excrete the eggs in their faeces, which can spread to their fur as they lick themselves clean. The minuscule, deadly cargo might be smeared around your home or could end up on your hand or under your fingernails when you stroke your pet. And from there, the parasites can make the leap to you, a dead-end host. The risk is even greater for cat owners, because catching mice is pretty much part of their job description. Unfortunately, fox tapeworm is also very happy to make itself at home in cat intestines, bringing millions of families into the at-risk circle. That's why pet owners should follow the official advice to deworm cats and dogs at least every two months using appropriate medication.

All things considered – including the risk of getting infected abroad or from your pets – the actual likelihood of infection when walking in the woods and picking berries and mushrooms would seem to be relatively low. The exception is certain at-risk groups such as hunters who bring home the foxes they shoot for taxidermy. Personally, I'm not going to rob myself of the occasional enjoyment of freshly picked berries. Looking to the future, the situation may become somewhat better for those of a cautious disposition, as more regions consider deworming strategies for foxes. This would likely involve bait being dispersed by plane, as with the rabies vaccination, with capsules containing a worming powder. According to the Technical University of Munich, the risk to people in one district of Bavaria could be reduced by 99 per cent, for example.[16] And yet I'm sceptical. If the fox tapeworm is really eradicated, what disease might take its place? And I would question whether we can permanently defeat the parasites. At our forester's lodge, we keep horses and goats, which we also have to deworm regularly. I know there's no such thing as completely worm-free animals; you can only reduce the infestation with regularly applied medication. Some worms survive treatment, and the result is the same as everywhere that we combat pests using chemicals: they become resistant. Therefore, we need to keep changing the chemicals that we use, and yet the tapeworms will keep on going and keep on adapting.

To bring the fox tapeworm under widespread control would therefore mean constant treatment with regularly adapted preparations, only to lose ten or twenty years' progress every time they develop resistance to worming drugs and take us back to square one. Does this sort of intervention in wild animals really make sense if it has to be constant and ubiquitous to be effective? Especially when we consider that there are simple ways to take precautions. And even those who (like me) don't want to forego uncooked wild berries can pay to have their blood tested regularly for antibodies to show up the presence of fox tapeworm, just as you can for Lyme disease. Perhaps it would even have been better to have tolerated rabies in the wild fox population. This may sound crazy, but how great would the danger actually be? Pets are vaccinated today anyway, so in a domestic environment the risks would be minimal. The ongoing presence of rabies would regulate the fox population, so they would rarely have any contact at all with humans. After all, you can see a fox approaching much better than you can see fox tapeworm eggs. And above all, you notice when you get bitten. Then you would go straight to the doctor, have an antidote administered by injection, and that's it. The hidden danger is perhaps the greater threat.

9. Little Red Riding Hood Sends Her Regards

THERE ARE WOLVES IN CENTRAL EUROPE AGAIN, thank goodness! Because as the old Russian proverb goes: where the wolf lives, the forest grows. Of course, wolves can't plant trees, but they can prevent too many from being eaten. The buds of the small saplings end up in the stomachs of hungry deer, which have rapidly increased in numbers, especially because of winter feeding by gamekeepers, as we've seen. Large populations of wild boar are also looting the forest floor and sniffing out almost all the beechnuts and acorns, leaving very few to germinate in the spring. It's the deciduous trees that struggle the most to produce offspring, and yet the primordial woodlands in our latitudes were almost exclusively deciduous. Where everything has been devoured, desperate forest owners plant spruce and pine trees, which the deer avoid like cattle avoid stinging nettles and thistles in the field. The needles are too prickly to chew, the resins and essential oils are bitter and sticky, as we've already mentioned, so the deer pass these by, looking for something more appetising. The consequence is that many of our forests have been transformed into monotonous evergreen plantations. So now the wolves are making a return to the wild, might we return to a level playing field? If the wolves are well, the forest is well – right? Of course, it's not quite that straightforward. Wolves eat meat, preferably from wild boar or deer, so naturally the numbers of these decline wherever a pack of grey hunters shows up.

The ways of nature are never that simple, of course. To get familiar with how the populations of predators and prey influence each other, let us first take a look in the garden. Here we have a classic competitive situation: we humans want to grow vegetables or beautiful rose bushes, but the insects, mice, slugs and snails in the garden all have their own interests in these well-tended and well-fertilised plants. If we don't want to move straight on to chemical warfare, we might want to think about attracting 'beneficial animals'. Whether it's ladybirds or great tits, hedgehogs or buzzards, these species are all just waiting obediently to help us control the pests in our garden. Really? Can we genuinely increase the number of predators in order to eradicate garden pests? Sadly, no. It's time to throw out the cliché of 'beneficial animals' because one species can never gain complete control over the other; the predators can only multiply if there is a continual stock of 'pests' to prey on. By then it's already too late for your garden, because until the next generation of predators is ready to intervene, your cabbage patch has had it. So, does it work the other way around? I learned when I was a forestry apprentice that the prey regulates the predators, not vice versa. That sounds logical, but it also doesn't accurately represent reality. Nature is more complicated than either simplified model shows, and in the case of an interplay of two species, complex fluctuations can be observed in both populations.

Let's take a look at the example of Isle Royale National Park, an island reserve in Lake Superior, Michigan. Here, nature has been carrying out its own unique experiment, which researchers have been observing since 1958. Moose first reached the island when the lake was frozen over, and in the absence of predators, they multiplied there magnificently. They munched their way through the undergrowth, destroying many of the young trees. But in another harsh winter, a wolf pack followed and feasted on the resident moose population. For the onlooking researchers, the island's physical isolation was a gift: both populations were virtually trapped, allowing for a controlled investigation into how the populations interacted and influenced one another in a relatively small space (well, the island spans over 500 square kilometres).

One might assume that as the number of wolves increases, the number of moose would decrease, as more are caught and eaten. Then the number of wolves would fall again, since they have fewer moose to eat, or they have to search and hunt for longer before they do catch one, meaning that more would starve to death. Then the moose population can rise again. But another hypothesis sees it very differently: when there's enough vegetation to support a larger population of moose, they will multiply, so that the wolves can also feed well. Indeed, the data has also shown that the moose's rate of reproduction increases the more their peers are killed by wolves. Conversely, a larger wolf population means more pressure on the predators because they now have to fight harder to defend their territories from each other. Fluctuations in the moose population therefore depend more on their habitat and on the availability of food than on the wolves, unless there is a particularly tough year. When there's a severe winter, food is scarce and many animals starve to death. If the rest are now pursued heavily by the wolves, there will likely be another slump followed by a sharp fluctuation downwards.[17] Are you now completely confused? I wouldn't be surprised. I'm telling you this to show that the way things are interrelated in nature is often not as straightforward as we were led to believe in the classroom. How then can it be squared with the old Russian saying about the wolf and the forest?

The answer is within sight if only we look at things differently. If we focus less on the fluctuations in the herbivore population and more on their behavioural changes, we get to the bottom of it. For this, let's hop over to Yellowstone National Park. This reserve also had far more herbivores than the habitat could support, mostly wapiti (elk). These large deer massively reduced the tree stocks and left entire landscapes desolate. The rangers further aggravated the problem by feeding the animals in the winter, causing their population to swell further. The turning point came in 1995, when the rangers, in collaboration with researchers, released wolves into the national park. By 1996, there was a resident population of thirty-one wolves, diligently knuckling down to their tasks of procreating and feasting, mainly on elk and smaller deer varieties. The wapiti population dropped steadily from 16,791 in 1995 to 8,335 in 2004, before stabilising at a lower level still. Meanwhile, the number of wolves rose to around 300.[18]

But more significant than the decline in their number, was the way that the herbivores adapted their behaviour. In the past, deer and elk used to graze on the river banks, ravaging the plants that formed a natural erosion barrier. Rivers and streams cut into the banks, washing away valuable soil. The muddy water suffocated the fish and other aquatic life, reducing the biodiversity to little more than just large and small varieties of deer. When the wolves were reintroduced, the deer and elk avoided the river banks, as they were more exposed there and at greater risk of being caught by the wolves. This meant the bushes and trees started growing back along the water's edge, shoring up the bank against erosion. As the trees matured, beavers could return as there were now enough trunks for building dams and branches to eat. Natural barriers began to force the rivers and streams back into meandering twists and turns, slowing the stream and reducing the rate of erosion. All of these changes to the ecosystem were triggered by the presence of a large predator, and a similar impact is conceivable in Europe too where wolves are being reintroduced.

Returning to the Little Red Riding Hood trauma, should we consider what risk the reintroduction of wolves poses to humans walking in the woods? What about in our towns? There was a furore in the German media in 2014, when it was reported that a wolf had been spotted approaching two children at a bus stop in rural Saxony.[19] If they hadn't got on the bus in time, the unspeakable could have happened – as Gerd Steinberg argued, founder of Germany's anti-wolf lobby group *Bündnis gegen den Wolf* ('Alliance Against the Wolf'). How can we keep our children safe from harm with wolves on the prowl on our streets and around our bus stops? We're back in Little Red Riding Hood territory, several centuries later. Just as the Brothers Grimm captured ancient fears and myths in their fairy tales, so too do modern storytellers. With urban myth quickly morphing into common knowledge (a friend of a friend heard about it, so it must be true), the imagination blossoms with such bizarre flowers that it becomes almost impossible to believe anything.

To shed some light into the dark thicket of the debate over wolves, we need to look at the various warring camps. First there are the gamekeepers and the deer hunters. The return of wolves to Europe undermines one of the key arguments they rely on to justify hunting: since there are no longer any large predators in the wild, human hunters should take on the role of controlling population levels. If the deer population is allowed to get out of hand, the meadows and forests would be ravaged. But besides thwarting their justification for their existence, there's another reason why wolves are a nuisance to the hunting community. How would a gamekeeper feel if a lupine rival casually mauled one of the roebucks he had been carefully nurturing for years, leaving the deer's magnificent horns to rot in the bushes? How would hunters feel about saying goodbye to the days of plenty, when they could sit watch in their high seats, virtually guaranteed a sighting of a deer or boar? What would become of the game hunters' weekends if they could no longer rely on spotting anything worth shooting? The gamekeepers' ability to plan and control 'their' stocks was at threat.

The wolf is therefore seen as the hunting community's rival, and they are fighting its comeback with every means permissible in law. Shooting the controversial beasts is an effective measure, and is indeed regularly carried out, but it is a criminal offence and there have been convictions. What is permissible, however, is influencing public opinion. And this happened in a way that was, at least initially, very subtle. The anti-wolf lobby claimed, for example, that wolves were being resettled. The truth, on the other hand, is that their natural return was allowed. The difference sounds marginal, but in fact it is significant. After all, the first could amount to an unlawful interference in the balance of nature, while the second can be seen as recognising national and international protection rights for a rare animal species that is now on the move.

In fact, hunters needn't worry quite so much, because as we have seen with the North American wolves, there is really no grounds for fearing that their return to the wild will spell the end for the deer. Well, some deer hunters might pass an evening or two without a sighting of their targets, but it won't mean forests are completely bereft of deer, otherwise the wolves would themselves starve to death. There are exceptions, of course. Artificially introduced animals such as mouflon sheep have nothing to laugh about in areas where wolves get re-established. These sheep have impressive spiralled horns and as such are sought after among trophy hunters, but there's a major drawback. They were originally introduced to the Mediterranean thousands of years ago as domestic livestock, but have long since gone feral. Because of their elegant head attire, they were introduced by gamekeepers to the woodlands of German-speaking countries, to boost the supply of sought-after trophies.

But in their adopted home, these mountain dwellers suffer for their beauty. Their hooves have evolved for clambering on rocks and grow fast for a habitat where they are constantly being filed down. On the soft forest floors, on the other hand, their 'nails' don't get filed down, so they grow longer and longer until their in-growing toenails start to fester. Many of these beleaguered sheep can only limp painfully and have little chance of evading a wolf. This efficient predator will opt for the easiest prey and sets his sights on the plodding mouflon. Where the two species have met, the mouflon have quickly disappeared from the wild. One might also describe it as the wolf restoring natural conditions. It goes without saying that this is a source of great displeasure for the gamekeepers, though. We've mentioned that the mouflon was artificially introduced, but isn't that true of the wolf, too? The only flaw in that argument is that it can't be proven; the grey hunter appears to have come back to Western Europe of its own volition.

The argument is still too good to be dropped, so now the hunting lobby keeps returning to it when they're grasping at straws. The *Jägermagazin* ('Hunter Magazine'), for example, ran a story in 2014 claiming that the German Federal Police had stopped and searched a lorry on the German–Polish border, which was – purportedly – illegally importing wolves and lynxes for release into the wild in Germany. The police press office had to release an official statement to set the story straight. Apparently, a vehicle had indeed been searched and was found to contain a 'Steppenwolf'. However, this was not an animal, but a stolen Steppenwolf bicycle that was being not imported but exported from Germany, where this manufacturer is based.[20]

The second group of wolf opponents are sheep farmers. They fear for their animals, which are out in the field waiting for predators like dinner on a plate. Eighty-centimetre-high fences don't do much to hold the grey hunters back from their buffet – if the animals are fenced in at all. It not only makes sense for nomadic shepherds to keep sheep in this way; in some places it's a tradition based on convenience. In Norway, for example, I've often seen how herds graze in the wild, left completely to their own devices, before being driven back to the farm in autumn to winter undercover.

A wolf will expend as little energy as it needs to become full, and a sheep is much slower than a deer or a boar. It's little wonder that the Norwegians tolerate only a minuscule population of wolves in the region of the border with Sweden, so they can safely round up as many of their sheep as possible after the grazing season.

Elsewhere in Europe, it is more common to keep sheep in fenced pastures. Many sheep and goat owners have relatively small herds, kept more as a hobby. Our three goats are not really part of our livelihood, but are more like pets. Protecting them from wolves was quite straightforward: we bought a higher electric fence. Most of the information we came across suggested ninety centimetres should be tall enough. Just to be on the safe side, we opted for a fence that is 1.2 metres in height. It's made of sturdy electric netting that even foxes can't get through. The essential thing is that the grass is always well mowed along the line of the fence. If the grass is long enough to touch the live wires then it directs the electrical current into the ground, rendering the fence ineffective. When it's working, the device sends a pulse through the wire fence every second. If you touch the fence, the pulse flows through you and it hurts, as I have discovered when I've forgotten to switch it off before working near it. The pain is so intense that I find I'm on my guard for weeks afterwards. The same goes for the goats inside the fence, but also for potential wolves on the outside. A fladry line could also be used as a deterrent – a strip of ribbons that flap in the wind, reducing the chances of a wolf attempting to jump over.

Now, professional sheep farmers will argue that this is too much outlay, and I can quite understand that for sizeable herds it could be prohibitively expensive. But in Germany, at least, the state has responded by offering funding. There are grants not only for the fences, but also to invest in special livestock guardian dogs. These large breeds live together with the sheep and may even believe that they are one of them. They stay with 'their' flock day and night, protecting them against any attack. Just making themselves seen is usually enough to scare off any wolves, or even a hiker who comes too close to the sheep. In contrast to lively herding sheepdogs, which run around all day long, guiding the flock to where the shepherd indicates with his whistle, these livestock guardian dogs doze near the grazing sheep and are barely noticeable at first sight. And yet, wherever these guardian dogs are on watch, there seems to be a peaceful coexistence between humans and wolves.

Now we've been talking about the wolf all this time without really introducing him. And yet it is important to understand how these beasts fit into our shared living space. For a start, wolves are predators; that is, they hunt other animals. Their diet includes deer and wild boar; humans really don't interest them in this respect. If they can't find any large mammals, then it could be smaller animals such as mice – even the tiniest snack can keep hunger at bay.

Wolves generally do not pose a threat to people; they tend to stay shyly in the shadows. They don't see us as providers of food or care, they simply keep their distance from us. But they don't avoid our environment, our 'urbanised landscape'. How could they avoid it? There's little else in mainland Europe, where there is barely an acre left of true wilderness. The primeval forests have been uprooted and large parts of the landscape have been transformed into artificial steppes of pasture and cereal crops. The landscape is carved up by roads and paths into tiny plots. There are around 650,000 kilometres of paved roads in Germany alone,[21] with another 1.4 million kilometres of surfaced tracks through forests for timber transporters. That doesn't leave a lot of space for peace and seclusion.

Wolves don't care about the roads and the traffic, as long as they find a safe, secluded spot to rear their cubs. What could be better suited for this than an army training ground? These private territories are certainly not quiet all of the time, but at least there aren't joggers and mushroom pickers poking around. There needs to be a sufficiently large supply of potential prey such as deer and wild boar – also not a problem, as we've already discussed. They seem to be preferred over domesticated sheep and goats, tamed by enclosure and breeding, which you would think would be easier to hunt. In the East German province of Lusatia – the region of Germany with the longest experience of wolves – researchers have been able to analyse wolves' diet by studying faecal samples. Roe deer make up the main part of their diet at over 50 per cent, followed by wild boar, red deer and fallow deer. Pets and mice were grouped together in the study because they make up a tiny proportion of the wolves' diet, at less than 1 per cent.[22]

Although wolves can be seen as a synanthropes – wild animals that live alongside humans and find their way into our cultivated landscape – they rarely attack our livestock. And the wild animals which hunters like to view as part and parcel of their licensed game-hunting reserves are in fact legally not owned by anyone; they belong to nobody but themselves. The only people who have genuine legal grounds for grievance are the few livestock owners who do not take up the offers of assistance to protect their animals. Otherwise there is a peaceful coexistence, which considerably reduces the case to be made against the returning wolves.

When all other arguments fail, the anti-wolf lobby falls back on the vague word 'problem'. Whenever it's placed in front of the animal in question, it has the power to attract the attention of authorities. Bruno the bear was an earlier example of this public relations tactic. This wild brown bear, officially named JJ1, was found rambling around Bavaria in 2006 and was initially given a warm welcome. But the locals were not prepared for the arrival of such a big omnivore. Just as with sheep farms in areas where wolves now roam and where farmers have had to adapt their defences accordingly, Bruno found unsecured beehives and helped himself to the occasional unsuspecting sheep. He took what he found before him and was labelled a 'problem bear'. The solution was rashly decided on: shoot it! Bruno is still to be found in Bavaria, but only as a stuffed specimen in Munich's Museum of Mankind and Nature. Isn't that the solution for wolves, the lobbyists cry. Haven't we already had enough risky close encounters?

Indeed, there are a lot of dangerous attacks by wolves in Germany – or by almost-wolves. The almost-wolves I have in mind are domestic dogs, which differ from their wild relatives in essentially one aspect only: they are tame, so have no fear of contact with humans and do the opposite of keeping themselves to themselves. The German medical journal *Deutsches Ärzteblatt* calculates that of the approximately 30,000 to 50,000 bite injuries in people registered annually, 60 to 80 per cent are caused by domestic dogs.[23] What if there were even ten wolf bites per year? I can imagine there would be massive protests in the region in question, fairly swiftly followed by action to kill these 'problematic' animals. But there hasn't been a single case of a person being bitten by wolves in recent years. And yet nobody wants to do anything to rid ourselves of the menace of dogs. Are we not guilty of double standards? In any case, I would be ten times happier to encounter a wild, free wolf in the forest than a stray German Shepherd dog – the latter would be far likelier to bite.

But if you really are lucky enough to one day find yourself face-to-face with a wolf, then your pulse is sure to go up. It might be good to have some tips up your sleeve. The best thing to do is stand tall, clap your hands and shout loudly to draw attention to yourself. You also need to stare the wolf straight in the eyes. If you feel unsafe, retreat slowly. The emphasis is on slowly, because running away can trigger the predatory reflex. Whatever you do, don't throw stones or sticks at a wolf, because that's just going to make it curious. To be on the safe side, you could also carry pepper spray – that's all you need. Wolves are just curious, not aggressive. In most cases the likeliest outcome is that you catch a glimpse of one from afar before it disappears again.

Incidentally, these tips for how to behave are from Elli H. Radinger, wolf researcher and author. She has summarised her advice together with the expert Günther Bloch in the book *The Wolf Is Back*. You should only take advice from a genuine expert. Unfortunately, there are more and more self-proclaimed, and officially recognised, wolf consultants who in fact have a very limited understanding of their behaviour, because they have had little opportunity to observe them in the wild. This is how these stories about 'problem wolves' can do the rounds so quickly; wolves which allegedly strayed unnaturally close to towns or villages, and dispensed with their usual shyness. Wolves are not interested in how we interpret their way of life, as long as they themselves feel safe and undisturbed. Elli Radinger tells us how a she-wolf was seen walking around with her eleven puppies in a mown cornfield near the Spanish town of Léon, while the farmer went on mowing with his tractor across the field. In the Romanian city of Braşov, another she-wolf was regularly seen – called Timisch – and was often taken for a dog because of her tracking collar. Nothing untoward happened there. We can learn from these examples a calmer approach to living in proximity to wolves.[24]

The fact that wolves fundamentally do not pose a threat should not, on the other hand, lead to a glorified view of them. It's not for nothing that it is illegal to keep one as a domestic pet, with a few rare exceptions, and the same applies to wolf-dog hybrids. After all, they remain what they are: wild animals which are not suitable for cuddling up with on the sofa. It is equally illegal to feed wolves, and herein lies the only real danger that can come from an excessive fondness for the animals. Feeding them could tempt wolves to overcome their innate shyness and to come closer than the minimum distance which otherwise prevents them from approaching us. This argument is of course immediately seized on by wolf opponents. They see this minimum distance being crossed in encounters with young wolves, which in very rare cases follow their curiosity and come within a few metres of people, only to then run away again. The really dangerous thing is that the fear stirred by these horror stories stays with us as a tingle in the back of the neck, just as with sharks. Since *Jaws* forever cemented the shark's reputation as a monster of the deep, countless nature documentary makers have tried to correct this misconception – so far, in vain.

The wolf is not the only large predator that is set to reclaim our forests. I've already mentioned Bruno the brown bear, the first local representative of his kind after many decades of absence. It's still open to debate whether a coexistence with the omnivorous bear is as unproblematic as it is with wolves. Bruno was presumably never going after our flesh, but since he also eats plants, his prey spectrum is much more consistent with our own. Bears will gladly accept whatever's on offer, be it agricultural crops, berries, mushrooms, honey – or small pets. But individual animals can get a taste for specific delicacies, if you want to call it that. A colleague told me a story from his time as an apprentice in Norway: there were bears in the mountains that had developed a taste for sheep udders. They went after them not for the milk; no, it was the delicate flesh that they wanted. They would swipe the sheep round the head with a heavy paw, then take a hefty bite from the coveted part of their stunned victims. It is quite understandable that the shepherds were outraged at the sight of these severely injured animals. Given the traditionally rather hands-off grazing style that I've already described, this resulted in a very low tolerance for the bears. Also from Romania there are reports about brown bears that have taken to roaming into the city centre. They like nothing better than foraging through bins since they've taken a fancy to human food. And that's what sets them apart from wolves. The latter seek out living animal prey, which is more likely to be found in the sparsely popu- lated countryside.

Whether it will be possible to reinstate bears as local residents across Central Europe will essentially depend on how we manage to keep the animals away from our populated areas. Scaring them away with noise and rubber bullets would be the most harmless remedy, but may prove ineffective. What is completely unneces- sary in the wolf's cause could prove to be a bitter necessity when it comes to bears: culling. In Sweden, which ranks alongside Romania as the EU's country with the largest brown bear population, this approach clearly functions well as sightings are extremely rare. In Norway, only about thirty bears roam the forests, meanwhile their neighbours in Sweden are hosts to more like two to three thousand. I remember how thrilled I was when I came across some footprints and a big pile of excrement on a long hike in the middle of the bear region. And that is exactly the point: while so many people focus only on the dangers, it makes more sense to appreciate the opportunities. Nature gets back a little biodiversity, and that increases the attraction of hiking, as the wolf has shown. Wolf tours are now popular not only in North America, such as in Yellowstone National Park, but also here in Germany. Predators who help prop up the tourism industry – that's the kind of head-lines I'd like to see.

We also mustn't forget about the wild cat the size of a German Shepherd which roams certain low moun-tain ranges across Europe: the lynx. However, they don't get by entirely without help, because they can't procreate as easily as wolves and therefore any illegal hunting has a particularly severe impact on their numbers. The justi-fication for hunting them is again their appetite for various types of deer, which makes them competitors for trophies in the eyes of some deer hunters. Since lynxes are shy, solitary animals that rarely leave the forest and keep an even greater distance from humans than wolves, you are unlikely to ever get to see these beautiful animals in the wild. You may, however, be lucky enough to spot some tracks in the snow.

10. Trees – As You've Never Known Them Before

I'VE NEVER BEEN A GREAT FAN OF GUIDED TOURS. Whether it's a walking tour around a city, a museum or out in the wild, unless the guide can inject a bit of humour and fun into it, then a lecture filled with facts and figures can be pretty dull, quite frankly. And because school groups often feel the same as I do, I make sure I don't fall into this trap when I'm taking them to explore the woodlands here at Hümmel.

Why should we only learn to recognise trees by the shape of their leaves and needles – and not by how they taste? The pupils from the local primary school have enjoyed putting this question to the test: the taste test. In spring, they had a little nibble on fresh spruce tips. These bright green buds are soft and have a gentle lemony flavour with a slight hint of resin. You can eat them raw, cook with them or brew them up as tea: all tasty ways to get to know the tree and recognise it as a species. This testing method caused a ripple of surprise at a springtime family activity day organised by the State Forestry Office. A few children were on an activity trail and spotted a spruce tree. Of course, they pulled off a spruce tip or two and took a bite to identify the tree. The manager of the Forestry Office, watching from the stand he was supervising, exclaimed with just a touch of disapproval, 'This must be one of Mr Wohlleben's groups!'

Clearly, I wouldn't advocate identifying everything by taste; there are poisonous plants too, as we shall see. But once a tree expert has helped you to recognise what you're looking at – be it a spruce, an oak or a willow – this new knowledge is so much better internalised when you experience it with all your senses, especially if you are a child. Sour yet juicy conifer shoots stick in the memory better than dusty old Latin names.

The same could be said about beech trees. In spring, their fresh leaves are delicate and ever so slightly bitter, without the resinous hint that spruce tips have. They're perfect in a forest salad. They don't store well so pick the leaves just before you prepare the salad, and add the dressing just before you eat, so that the leaves don't disintegrate. If you pick from the lower branches of a large tree, you won't harm it in the least. You're in good company, in fact: many species of beetles enjoy these tasty leaves, as do deer.

So if you fancy having a nibble through nature's culinary guide book, it's worth knowing that you can eat the shoots of many native tree species. Maple, birch, oak, lime, pine, larch or even fruit trees: all produce fresh shoots that taste good, and each with slightly different flavours. But when I said 'many', there are, of course, exceptions. Yew needles, for example, look very similar to those of the fir tree, but while fir tips are also edible, yew tips are highly poisonous.

Our noses also have a role to play. When you rub the needles of the Douglas fir between your fingers, they give off a scent like orange zest. Oak bark and wood have a strong aroma of tannin and indeed in the past tannic acid was extracted from the bark. It functions as a repellent and is the reason why oak garden benches, for example, are mould-resistant.

There is a non-native species (more likely to be found in parks and gardens) that stands out for its particularly bad smell. It is the ginkgo tree, an ancient relic from an evolutionary age, which is often described as a living fossil. The world's oldest living plant, the gingko or maidenhair tree is widely respected for its medicinal properties too: extracts from its leaves have long been used in homeopathy to treat all manner of conditions. But when the trees blossom, the smell is likely to make you feel nauseous. The female specimens produce fruits that reek of rancid butter, or vomit. So if you're considering getting a gingko to give shade in a large garden, it might be best to go for a male specimen.

Is this discussion of aromas and flavours still a little too vague for your taste? Would you like a more detailed description of the most common species? Let's have a look at a few in turn.

Spruce: feeling the heat

The spruce, or more precisely the red spruce (*Picea abies*), has become one of our most common species. At least every fourth tree in Germany is a spruce, and I mean that literally, because they were usually planted deliberately. These trees prefer it damp and cold, so are most at home in the classic taiga climate, such as northern Scandinavia or the Alps. They're also now found in many low-lying regions, and their popularity among forest owners and rangers is chiefly due to two characteristics: they always grow straight up, in defiance of the earth's gravity, and they don't get ravaged by deer as the prickly needles put them off seeking out the shoots. The wood is suitable for building and for paper, making it a safe investment. There are, however, a number of good arguments against the cultivation of spruce. From an ecological point of view, there are thousands of small animal species that have no appetite for the bitter needles. Since this leaves them with very little to eat in the dark spruce plantations, they die out locally.

Spruce can be distinguished from other types of conifers by its bark (smoother, reddish brown) and the cones (often at least ten centimetres in length, light brown), making it easy to identify from the ground.

Spruce is set to disappear from the forests of much of Central Europe in the next few decades, as global warming heralds a drier, hotter climate. In many regions, that spells the end for the cold-loving northerner, including where I grew up in Sinzig on the River Rhine. The climate in the valley is almost Mediterranean, with temperatures often a good four degrees above those not so far away in the Eifel region, where I now live and work. Every year sees an infestation of the European spruce bark beetle, a tiny weevil which attacks evergreen plantations. It is a weak parasite and only bores into trees that can't properly defend themselves, such as when it's too dry and the trees can't produce enough resin to resist an infestation and drown the invasive beetle; it's as though they run out of spit. A rise in average temperatures by just one degree – which we're already seeing in many parts of Central Europe as climate change takes its toll – means the spruce is increasingly vulnerable. In Sinzig and many similar places, this combination of climate and pests spells the end for this conifer.

Pine: elegant, but trouble

The pine has experienced a similar triumphal procession through Europe as the spruce. Commercial forestry has taken this tree far beyond its natural range, which was also originally in the cold, damp north. According to the National Forest Inventory, pine makes up just under 25 per cent of the total area of German forests. I've got nothing against pine, per se; it's a beautiful tree. We have some around our forester's lodge that are around 140 years old. They're an elegant sight with their short cones, their long needles growing in a neat arrangement of pairs around the twigs, their deeply furrowed bark which at the top of the trunk becomes a smooth, orange-hued skin. Since the house dates from 1934, our pine trees must have already been there before it was built.

Where individual trees can be an asset to a garden, hundreds of thousands of them can form a green desert. This is already the case in Brandenburg, for example, where hardly anything else can grow among the pines planted in monotonous rows. Forest fires, once completely unknown in Central Europe's deciduous forests, are now a serious hazard, requiring considerable effort and outlay to keep them under control. Pine is incidentally at the bottom of the list of types of wood in demand, which is another good reason to finally bring its steady march across Europe to a halt.

Silver fir: a conifer that thinks it's deciduous

Sometimes you're in the woods and you hear someone shout, 'Look, there are some fir cones on the ground!' There's lots of things they might be, but they won't be fir cones, because these crumble while they're still on the tree. No cones on the ground – that's the first sign of a silver fir. The second indicator is the way the flat needles grow to the sides, as though the twig had a centre parting. On the underside, the needles have two white stripes lengthways. They're not prickly, and they are also darker in colour than spruce needles, which have a slight yellow tinge. If the tree you're looking at also has a silvery-grey bark which is quite smooth like the spruce, then it's likely to be a silver fir.

Silver fir is the most deciduous-like of the conifers. It tends to grow well in mixed forests together with beech trees, so it is occasionally to be found amidst ancient beech forests. Its roots stretch deep down, and its needles are mild tasting and a popular snack for animals on the ground; in this respect, it would be tempting to categorise the silver fir among the deciduous trees, if it weren't for its evergreen needles. It doesn't (yet) appear naturally in the north of Germany, as it was one of the last tree species to have returned here after the Ice Age. How did that happen? Perhaps it was through birds carrying the seeds gradually further north. The spotted nutcracker, for example, dutifully stashes away fir seeds for the winter, burying them in the ground sometimes several miles to the north of the mother tree. The seeds germinate when the nutcrackers don't manage to eat them all, and the fir is able to get established further and further north. In theory, anyway. What sets it back is that in contrast to deciduous forest birds such as the Eurasian jay, the nutcracker seeks out particularly dry patches of ground, to ensure the winter supplies last better. The surplus seeds then tend not to get enough water in spring to germinate. Bad luck for the tree, but good planning on the nutcracker's part.

Silver fir saplings are also one of the favourite delicacies enjoyed by deer, so the tree has disappeared in many places due to large herbivore populations.

Common beech: mother of the forest

I didn't coin the emotive name 'mother of the forest'; no, it has been used by foresters for generations. Why has this tree in particular earned such a special title? It may well have something to do with their astonishing nurturing qualities. The old mother trees shelter their offspring, which grow little more than a metre in a hundred years in the dim half-light. This slow growth is necessary if the young beech is to live to a grand old age and not exhaust its strength prematurely. So that the young trees don't starve for lack of light, they are literally suckled by their mother tree, which pumps a sugar solution to her young through their root network. The mature trees are equally attentive to each other once fully grown. They help their weaker family members with similar nutritional aid, and are supported by the others in the event of illness. The result is a robust community that together is much more resilient than a single beech on its own.

And yet the ancient forests are in grave danger. The beech was once the German tree and occupied approximately 80 per cent of the area of our impressive woodlands. A great proportion have been cut down, felled to make way for fields and cattle pastures, some of which were later partially reforested. Unfortunately, spruces, pines and other conifers have often been planted – and continue to be planted – instead of our native beech and we are still a long way from a return to our natural ecosystem. A minuscule percentage of halfway intact ancient beech forests still exist and lamentably most still do not have any protected status.

Oak: down to second place

What's happening to the German oak? The tree is rarely out of the headlines. In the central Rhine-Main region and many other forested regions it has been because of their crowns dying off. In urban settings, it's the fatal march of the oak processionary moth, with their allergenic hairy caterpillars. Among woodland competitors, it often comes up against the beech, so that it risks disappearing in many places unless there is human intervention. And yet the oak is a symbol of unshakeable stability and perseverance. Is that based on pure myth? Not quite. In the past, the oak tree bore much greater significance and value to humans than it does today. Not only buildings, but fleets of naval vessels were crafted from its tough wood. The autumnal acorn harvest was a blessing that helped to fatten up the swine before slaughter. The fruit of woodland trees is called 'mast', and a year with a bumper crop of acorns and beechnuts is called a mast year. Meanwhile, German has a related verb: *mästen* means to fatten an animal up.

And today? Oak trees would not naturally form entire forests in Germany, but tend to occur as individual specimens. When oak is grown commercially, though, rangers prefer to plant entire oak plantations. This carries with it the same risks as with other plantations. Oak-loving caterpillars have been known to strip whole forests bare, and it is all too easy for the dreaded oak processionary moth to spread when there is an abundance of oak trees in close proximity. They need trees with sunny crowns, and they are to be found in the forests everywhere in Germany: regular thinning leaves gaping holes in oak forests through which sunlight can fall, making a warm, cosy home for these tiny pests.

Birch: meaner than it looks!

If it's got white bark with black marks, it's unmistakably a birch. Strictly speaking, there are two types of birch tree you might encounter during forest walks in Germany: the downy birch and the silver birch. But since the first is very rare, we'll focus on her much more widespread sister, whose Latin name *Betula pendula* refers to the thin, limp branches which dangle down pendulously. We humans often associate drooping with sadness, because when we humans adopt such a pose it suggests a lack of driving force. (The weeping willow is a rather exaggerated example.) But if our estimation of a silver birch is limited to this analogy, then we completely underestimate its strength. Its intentionally floppy branches are actually more of a secret weapon than a sign of weakness; like a dominatrix's whip, they flick back and forth in the breeze, beating its forest neighbours out of its way.

We have a large Douglas fir in our garden, a conifer from North America, planted by my predecessor. Standing at about thirty metres tall, it has a large crown with soft, bluey-green needles. Beside the Douglas fir stands an eight-year-old silver birch, which can no longer keep up with its larger neighbour. Birch trees are sprinters which grow astonishingly quickly in their youth and tend to overexert themselves. By the age of thirty, they begin to decline and are overtaken in height by other species. Losing the race for height is dangerous for trees, as it means being overshadowed by their neighbours. In the shade, trees are less able to photo-synthesise, which in turn means starvation, years of illness and a fatal decline after just a few decades.

The Douglas fir at the forester's lodge has over-taken the birch next to it, but the birch is not going to give up easily. Its branches hang long and slack, but in the wind their true nature is revealed. They swing back and forth, lashing at the branches of the Douglas fir. Is it too much to call this a targeted attack? If we have any doubt, we need only take a closer look at the bark of the birch branches: it's studded with warts, making the surface rough like sandpaper. Even a rock is worn down by constant drips, as they say, and the constantly swinging birch branches wear away the branches of the Douglas fir, making it shed needles on that side. Over the years, the Douglas fir's crown has been worn away to reveal the birch's outline. David has defended himself against Goliath, for a few decades, at least. The birch will eventually run out of steam and die of old age, leaving the Douglas fir to grow old in dignity.

Larch: a tree with no future

Yet more bad news? Is the larch also on the way out? It's not quite so bad, but we're gradually heading in that direction. A larch is by nature as rare a sight as the spruce, as it also hails from further north or the moun-tainous regions just below the tree line. The larch is a funny tree. While all other conifers remain beautifully green throughout winter, the larch turns golden together with the deciduous trees in autumn and then drops all of its needles. A dormant one might be taken for dead spruce, if you can't tell the difference. Sadly, I don't know why the larch does it, but it at least makes it easier to identify.

Even at agricultural college, our lecturers quoted the local dialect saying, '*Lärche auf der Bärche* (*Berge*)' ('larch in the mountains'), reminding us that this tree species loves the cool, damp highland climate. But as with other conifers, commercial forestry has meant that the larch has been mercilessly relocated to the lowlands for better timber yields. And because the European larch, as it is properly known, still didn't bring in the desired profits, the Japanese larch was imported for its much faster growth. Unfortunately, it turns out that the Japanese variety is very happy to breed with its European counterpart, producing a hybrid offspring. This is regrettable because pure-bred European larches are becoming ever rarer and may eventually become extinct. It's also getting harder to recognise one when you see it in the wild. The native species has yellowish branches and the cone has closely fitting scales. The imported variety, meanwhile, has reddish branches and the scales of the cones curve outwards; seen from above, they resemble rose petals. More and more hybrid forms are emerging, until in a not-too-distant future the distinct species may no longer survive.

Incidentally, wild apples and pears are suffering a similar fate as they have been diluted by the cultivated

varieties and are finally being genetically displaced altogether. After all, bees pollinate all the apple trees indis-
criminately on their blossom tours and therefore play a
role in cross-breeding. Whether or not there are any
purebred apple and pear trees left in the wild is a matter
of scientific debate.

Ash: victim of globalisation

The ash was very important to our ancestors. It played
a major role in Norse mythology as Yggdrasil, the
'world tree', with its immense canopy spanning the
heavens. It is easy to recognise. The European ash has
black, pointy buds and long compound leaves made up
of several leaflets; altogether, the compound leaf can
reach a length of forty centimetres. The feathery leaves
are easily confused with rowan, which is why the latter
is also known as mountain ash. The rowan has very
different buds, however, and it doesn't grow as tall.

The ash tree is currently under threat and the
aggressor is a tiny fungus known as *Hymenoscyphus
fraxineus* (previously as *Chalara fraxinea*), which attacks
the branches and causes them to die back. The bark
turns beige and the ash tree struggles to produce enough
food through photosynthesis. The tree dies slowly, over
many years. The jury is still out about whether the
fungus is a mutated variant of a native species or whether
it was imported. It seems to have come from the Asiatic
region, possibly from Japan, and may have arrived in
Europe via imported goods containers. The Chalara ash

dieback, as this epidemic is known, has gradually spread to every corner of the European continent, where it has carried off up to 90 per cent of ash trees. However, there is still hope for the ash as a species: the remaining healthy trees appear to be resistant to the fungus, and scientific studies suggest there's a good chance these healthy trees will reproduce, resulting in the emergence of healthy ash forests once again.

Incidentally, the ash is not the only victim of globalisation among the trees. The elm has suffered the same fate several times. In the early twentieth century, an aggressive sac fungus (Ascomycetes) was introduced with imports from Asia. Dutch elm disease is spread by elm bark beetles, which unknowingly transport the fungal spores with them, as they roam from tree to tree boring holes in the bark. The fungus clogs up the vascular system, blocking the flow of water and nutrients and resulting in the death of the tree.

From Europe, the epidemic hopped over to North America, before returning to us in an even more aggressive form. The result is that we now find elms only in remote locations as individual trees as yet untouched by the roaming elm beetles with their deadly fungal cargo. And unlike with the ash, tragically the survival rate in infested areas is zero.

11. Is It Really Love?

IMAGINE THE FOLLOWING SCENARIO: you've been walking for hours through the shady woods and it's time for a lunch break. All of a sudden you spot a small, grassy clearing in the warm sun. Wouldn't this be the perfect spot for a picnic? What's appealing about this clearing is that there are no trees here. Does this suggest we don't actually like the forest, but prefer sunny landscapes with just single, impressive trees? The question sounds heretical, but it is crucial to our relationship with nature. From a developmental point of view, humans come from the open steppe. We are ideally equipped for dry and hot climates. Our upright posture means that only a small area of our body is heated by the sun, and our relatively hairless bodies can cool very effectively with the help of sweat glands. These qualities enabled our ancestors to hunt by chasing their prey until it overheated and collapsed. A key factor in this was excellent vision, which helped hunters to locate animals from a distance. The other senses – hearing and smell – were neglected in comparison.

These characteristics are less useful in the woods. When the sunlight barely breaks through the dense canopy, an inbuilt cooling system is less important; better to be hairy and able to keep yourself warm. Forest animals are equipped with a completely different set of features. Eyesight is not so important, but a sensitive nose and big ears are. What good are eagle eyes when your long-distance gaze is blocked by tree trunks after just a few metres? Enemies can only be perceived in time if you can smell them from hundreds of metres away and hear their footsteps on cracking branches. And because larger groups quickly lose each other in under-growth, forest dwellers are typically solitary animals.

Our ancestors therefore found woodlands to be an ecosystem that only partially supported their needs. They kept the cold at bay with fur blankets and fire, and improved the visibility by clearing large swathes of forest. Felling the trees. By nature, humans are not comfortable living in the forest, and this is obvious if you look at the way we have cultivated our landscape: we've created a perfect artificial steppe with our fields of wheat, barley and maize – all grass species, albeit particularly productive ones. Fields of grass crops, meadows for cattle (also steppe dwellers) and just a few small patches of woodland here and there – even 200 years ago, this was the general picture across Central Europe. Much has been reforested since then, mainly prompted by the demand for wood as a raw material. Meanwhile, the deep, dark woods of yore were associ-ated with creepy fairy tales and ghost stories.

That negative association with forests is generally a thing of the past, don't you think? Think of the clearing I described at the start of this chapter – a light, airy spot where almost everyone feels comfortable. This is also the reason why foresters clear gaps in larger forest areas to open up the vista. These cleared areas tend to be at striking vantage points, and all that's needed to make the most of the feel-good factor is a bench to sit and enjoy the view. These rest points are very popular. It seems our age-old instincts influence us more strongly than we like to admit in this era of rational thinking. Our love of the forest may stem from another aspect of innate nostalgia: it is the last reasonably intact ecosystem we have.

We've been strolling among the trees for a while now and yet we still haven't asked ourselves the crucial question: what is a forest anyway? For the official answer, we can look to the legal definitions. The second section of the Federal Forest Act of Germany, for example, defines it as any territory where forest plants grow. However, timber storage areas, paths, small meadows and clear-cut plantations also belong within the defini-tion of a forest according to the law, provided they are surrounded by sufficiently large stands of trees. It quickly becomes clear that, in the eyes of the law, wood-lands are defined in purely economic terms. Why else would they have thought to include larger areas without any trees? Of course, it follows that even clear-felled plots and stretches of spruce ravaged by storm still belong to the forest, when there are no living trees on the land. The only requirement under the law is that they are reforested within five years. So, can we reduce the definition to one common denominator? How about any plot of land containing a closed-canopy group of trees – can we agree on that?

Perhaps visitors from abroad are best placed to judge whether our paltry clumps of trees can really be described as forests. Our perspective is perhaps a little clouded by emotion, so maybe outsiders can make a more neutral assessment. I once had the honour of hosting Dr Ali Ost Montazeri, Iran's chief forester, when he was visiting Germany in 2009. When we started to discuss our forest, he said dryly, 'Forest? What forest?' For him, what he had seen on his tour was merely a plantation landscape. Another conversation that comes to mind was during a trip to Gabon with my mother. She was very animated as she described the forests of Germany to a local. Imagine her surprise when he responded that he had indeed been to our neck of the woods. He had sought out the famous Black Forest and had wandered through the low mountain range, but to his disappointment he had been unable to find the great forest he was looking for. All he had found were conifer plantations.

'Hang on a minute,' you might say. Indeed, my German readers are likely to object quite indignantly. Aren't Germany's forests a haven of sustainability? An exemplary ecosystem that has been carefully cultivated for centuries? Aren't they a model which our development aid officials urge other nations to imitate? That's the version propagated by the authorities, at any rate.

But what is sustainability? Three hundred years ago, the idea of sustainable woodland management was that you shouldn't fell more trees than could grow back. It may seem obvious to us now, but it wasn't back then. The forests were plundered for charcoal and timber.

Charcoal was in high demand for smelting and for emerging industry. The forest would have been full of smoking charcoal piles – traditionally built kilns, with the journeymen who did the work living out in the wild alongside them. These skilled workmen layered the chopped wood into large heaps, which were then covered with soil and turf. Once lit, the whole thing smouldered for days, until the wood turned to black charcoal. This light fuel was much easier to transport to the ironworks, where it fuelled the smelting furnaces. Ironically, the significant spread of forested areas in the last 200 years is due to mining and the switch to coal as a main source of fuel. The discovery of this readily available fossil fuel rendered the much more work-intensive charcoal redundant – and the trees could recover.

As for the concept of sustainability, this was a term first used by Saxony's chief mining authority Hans Carl von Carlowitz in 1713, to express the idea that the quantity harvested and consumed should never exceed the amount that can potentially grow back. This was not a consideration motivated by ecology; no, in this era it was a question of safeguarding the supply of raw materials. After all, this is as much as a farmer can do: you can't reap more corn that you can grow.

We needed an updated definition of sustainability in the light of our current environmental problems, an issue that was tackled in 1992 at the UN's Rio de Janeiro Earth Summit environment conference. Since then, the sustainability of an ecosystem has been evaluated not only in quantitative terms, but also in terms of the quality and functionality of the environment, which should be passed on as fully as possible to our descendants. It remains questionable whether there has been success in implementing this goal of sustainability, because forestry in the German-speaking world is still very much fixated on the principles outlined 300 years ago by von Carlowitz. But thanks to our right to roam, there are a few ways we can check for ourselves to what extent we're succeeding with respect to sustainable management.

There are a few criteria that can help you identify whether it's a plantation or a semi-natural wood you're looking at. The most obvious starting point is, are the trees growing in neat rows? This doesn't happen naturally, so if they are it's the work of foresters with their Prussian love of discipline. Although it doesn't matter in principle how seedlings are planted, in a German forest everything has to be tidy and efficient. I learned in one of my early jobs that the first thing to do before planting a tree-free area is to set out the lines with ranging rods. These are the red-and-white-striped poles about two metres long, which are rammed into the ground in a line, to help you lay out the rows where the saplings are to be planted. And since the trees can't move from the spot, you see these same rows even after decades, unless the trees are heavily thinned for harvesting the timber.

The second question is the tree species. Unless you're at a high altitude in the mountains, approaching the tree line, then pure coniferous forests in our latitudes are always artificial plantations. We have already talked about the reasons for these plantations and touched on the problems. Beetles attack conifers which are less able to quench their thirst in these southern climes than they would be in their distant Nordic homeland, and evergreens are easily knocked over in storms. The resulting large bare patches that remain after the fallen giants have been cleared are typical of a plantation forest. Clear cutting is also carried out intentionally in order to harvest large plots efficiently. Unfortunately, it also happens in the last remaining ancient beech forests, which are then often replaced by North American Douglas fir.

But even when there are only deciduous trees, it doesn't necessarily mean the woods have emerged naturally. Teak or mahogany plantations will not replace a lost rainforest, and likewise beech or oak plantations are no substitute for the virgin forests Europe has lost. A woodland landscape is healthier when all age stages of offspring grow together beneath the canopy, born from the seeds of the parent trees. Even if now and then an old tree is felled (a process known as selection cutting), this diversity of ages among the plants brings the plantation closer to a natural ecosystem. It's rare to find very old trees and dead trunks in a *Plenterwald* – that is, a forest managed using selection-cutting techniques – so a good compromise is managed plantations interspersed with protected areas where older and dead trees are allowed to rot. Unfortunately, this kind of harmonious coexistence of commercial forestry and intact forest ecosystems tends to be found in less than 5 per cent of all woodlands, because the protection of forests is still not taken at all seriously despite legal provisions.

Apropos the protection of woods and forests, I'm afraid I can't spare you a rather unpleasant digression on this subject. It's not the forests that are unpleasant, but what the people who manage them do to them. For many decades, they have repeatedly sprayed lethal chemicals to eliminate one category of living things or another. An early climax in this tragedy was the use of a derivative of Agent Orange, the infamous defoliant that was used by the US military during the Vietnam War to kill entire primeval forests so that enemy soldiers had nowhere to hide beneath the bare crowns. Not unlike the herbicide-spraying campaigns in Asia, Central Europe was also flown over by helicopters spraying chemicals intended to wipe out the deciduous forests, which had gone out of fashion. Beech and oak trees held very little value at that time; low oil prices meant that no one was interested in firewood. The scales were tilted in favour of spruce – sought after by the timber industry and safe from being devoured by the high game populations. Over 5,000 square kilometres of deciduous woodlands was cleared just in my local region of Eifel and Hunsrück, through this merciless method of dropping death from the air. The carrier for the substance, sold under the trade name Tormona, was diesel oil. Elements of this mixture may still lurk in the soil of our forests today; the rusty diesel drums are certainly still lying around in some places.

Have things improved now? Not completely, because chemical sprays are still used, even if they're not directed at the trees themselves. The target of the helicopters

and trucks with their atomising nozzles is the insects
that feed on the trees and wood. Because the drab spruce
and pine monocultures give free rein to bark beetles
and butterfly caterpillars, these are then bumped off
with contact insecticides. The pesticides, with names
like Karate, are so lethal for three months that mere
contact spells the end for any unfortunate insects.

Parts of a forest that have been sprayed with pesti-
cide are usually marked and fenced off for a while, but
wood piles at the side of the track are often not considered dangerous. I would therefore advise against sitting
on them when you're ready for a rest stop and look out
for a mossy stump instead, which is guaranteed to be
harmless. This is quite apart from the fact that freshly
harvested softwood is often very resinous. The stains
don't come out in the normal wash; you need to attack
it with a special stain remover. Stacked wood carries
another danger: the whole pile is liable to come crashing
down. When you know that a single trunk can weigh
hundreds of kilograms, you tend to stay away from a
precariously stacked pile. It's not for nothing that the
German name for a wood stack is *Polter*, as in the
crashing and banging of a poltergeist.

Back to the poison. In areas sprayed by helicopter
I wouldn't pick berries or mushrooms for the rest of
the summer. Otherwise, the forest is low in harmful
substances compared to industrial agriculture. But does
it still count as a near-natural ecosystem? It ought to,
as German forestry is considered exemplary, after all,
with our approach to the use, protection and recovery

of woodlands being exported all over the world via our development aid agencies. If you take the state forestry administrators' word for it, then you'll assume our forests are growing, flourishing and prospering. Everything is living in beautiful harmony; humans and nature are at one again among the trees.

But is there really anything in it? I'm very wary about such claims these days, sceptical about such a positive spin. Not only because I've often witnessed something quite different when I'm out and about, but also because of the glaring conflict of interest, which even the Federal Cartel Office has started to notice. A number of state forestry commissions fulfil not only their official tasks, such as monitoring private forestry for compliance with the law; they also double up as market-dominating suppliers who sell the most timber and offer the largest range of services. And because their budgets are often topped up by the taxpayer, they can readily undercut the prices of private foresters, driving out competition in many places. It's more or less comparable to a situation in which the Inland Revenue was the largest provider of investment products. Who exactly is going to provide unbiased oversight? And that's not the only problem. How are you, a layman, supposed to distinguish between official communications and public relations? And PR and marketing can be mighty persuasive with its own special language, which I'm sure we've all fallen for at one time or another.

12. A Pocket Dictionary

EVERY PROFESSION HAS ITS OWN JARGON. It is not absolutely necessary; after all, most concepts can be expressed with more widely understood terms. And yet these charming old linguistic traditions reflect a culture we should seek to preserve, as long as the terminology doesn't obscure the facts.

I remember my first days as an apprentice in the Rhineland-Palatinate state forestry commission, when I had my first experience of this harmless variation in language. We were checking the damage after a storm and the forest ranger I was working with asked me to pass a *Kluppe*. To me that sounded funny, like he was asking for a clothes peg. I couldn't resist a grin.

'What is that exactly?' I asked, with the innocence of a newbie. The forester rolled his eyes and went over to the car himself. Out of the boot he pulled a calliper, a measuring tool, and passed it to me.

'For measuring the diameter of the trunks that have fallen along the road,' he muttered.

I've got nothing against traditions when it comes to technical jargon. On the contrary, they show that the craft in question draws on a long history, and that's particularly important in the forest: after all, the trees that we make use of today were planted generations ago. It seems rather sly, though, when these supposedly harmless synonyms are used to influence the public. One example is the German word *Waldpflege*, which literally can be taken as meaning 'forest care'. How might we interpret this word? Perhaps, if foresters care for their forest, the forest is better off as a result. The forest is fit and healthy, able to withstand pests and the challenges of climate change. That's what most people would understand by it. But what would you say if a butcher referred to his trade as 'animal care'? Bizarre? And rather sinister? And yet the term would communicate just as much as 'forest care' does for forestry. Forest maintenance, or 'forest care', in reality means little more than felling trees. It starts with the very young ones. Logically, the process is called 'nursery care' (*Jungbestandspflege*), and it involves dense growth being felled with the chainsaw. This leaves the remaining trees with more space and light to grow faster. This is called 'thinning' and it's the same process at later stages: the forester makes more space for the most attractive trees by chopping down their neighbours.

Another rather old-fashioned term for managing young stock, which is nevertheless still in use, is *Läuterung*, meaning 'purification' or 'refinement'. Speaking of 'purifying' the forest stand reminds me of the Middle Ages and purification by fire. Does the comparison seem a little harsh? Well, looking at it the other way, can we say this method of 'caring' for trees is actually good for the forest? Certainly not, and you can deduce that for yourself. Consider the Amazon rainforest. Does it really make it a healthier ecosystem when trees are felled to leave tidy, empty spaces? Of course not, and it's the same with our woodlands. When trees are sawn down, it always and without exception weakens the surviving trees. It already happens naturally with storm damage. Where once a tree was protected from the wind by its neighbour, this gaping space now leaves it exposed to the elements. It can take at least three years before the trunk, crown and root system adapt to the new danger. The other impact is the loss of the trees' social network. This is particularly evident in old deciduous forests, where sociable trees that have been reduced to lone specimens by logging are visibly ill. The tips of branches in the crown die back, leaving beech and oak trees with a bald, plucked look. A forest is not made healthier by felling trees, and this approach to forest management shouldn't be called 'forest care'.

What if we adopted the more honest terms used in animal husbandry and talked about 'tree slaughter'? Does it sound too brutal? I would welcome a change in language in this direction, because it would make it clear that it means dispatching sentient beings from life to death. While there is nothing wrong with this in principle, it might perhaps make us think twice about using wood. I'm sure I'm not alone in thinking this: it seems to me that many foresters feel uncomfortable about what they do as part of their daily work. But we sleep better when we couch the whole thing in comforting language that insulates us from a guilty conscience and external criticism.

The mention of crown dieback leads us to the next term: 'forest dieback'. You might think this term is well known and that there's no need for a translation, and I'd respond with a clear 'yes and no'. Forest dieback, which began to be reported in the media in the early 1980s, doesn't have a secret underlying meaning. It refers to severe damage to needles and leaves, and to entire mountain ranges of dying trees: the impact of acidic exhaust fumes from industry, households and traffic.

What followed was a real success story of environmental policy. The introduction of desulphurisation plants and catalytic converters prompted a turnaround and massive reduction in pollutants, with the result that the most threatening form of forest dieback seems to have disappeared. There are still problems with emissions, though now the focus is more on nitrogen oxides from agriculture and transport. It is not just the build-up of acids that is the problem, it's also the fertiliser effect that makes trees grow 30 per cent faster than they would normally. This boosted growth means the yield charts used by foresters to estimate the annual growth in timber quantities are no longer reliable and need to be constantly adjusted upwards. But it is no reason to celebrate. A greater volume of timber ought to mean more cash in the coffers, and to begin with that is the case. With time, however, the rapid growth means the trees quickly run out of steam, if you like; they become exhausted and are more vulnerable in the face of disease and drought. We mustn't be complacent;

there's still a long way to go in the campaign for cleaner air.

But to return to my 'yes and no'. It isn't as straight-forward as it seems at first sight, because at present the understanding of 'forest dieback' is coloured by how it is presented to the public, which is with certain facts concealed. Scientists agree that our woodlands are in good condition overall. We're not seeing extensive dieback, the lumber yields are encouraging, and, despite fertilisation by nitrogen oxides, the ecosystem is not in danger.

This is contradicted by the annual Forest Condition Report produced by the Federal Ministry of Food and Agriculture. It describes the crown condition of all forest tree species, which is no less worrying than in the past. Less than half of the trees are classified as healthy, while the majority are classified as moderately or severely damaged. This doesn't suggest the turnaround that has been spoken of. The forest is basically healthy, and yet most of the trees are sick? How does that work?

It is the forestry industry itself that is still to blame for impairing the health of the trees. The loss of the trees' social structures, as mentioned above, is only one aspect. Add to that the changing internal climate of the forest, which is becoming drier and warmer due to increased exposure to sunlight. The damage caused by the harvesters is particularly depressing – quite literally. These heavy machines compact the soil, crush the roots and spread fungal diseases, all of which make life diffi-cult for the trees. All of this is evident in the crowns, which are quite rightly classified as damaged in the annual surveys. The problem is the surveys are carried out by the same bureaucratic apparatus that is also the country's largest forest management authority. It's little wonder that they prefer to point the finger of blame at other polluters.

13. The Lumberjack Lads

IN THE PAST, LIFE WAS TOUGH, but the pace was a little more leisurely. The forester was enthroned in his forester's lodge, and every Sunday a flock of forest workers would come streaming in for their meagre wages. Spruce, pine and beech were felled by hand from late autumn. The only tools were axes and two-man saws, with which the workers would labour away at the partially frozen trunks until the trees fell. Then came the bark spud, or peeling iron: it was an exhausting job to scrape off the bark to reveal the bare timber, which could then be pulled by horses to the nearest track. Because the tasks were slow and laborious, there was plenty of work to go around. A large part of the male rural population was involved; in fact, forestry work was mainly carried out by smallholders who wanted some income in the winter months when there was less to do in the field.

Chainsaws entered the picture in the 1950s. These new contraptions were such a novelty that the old schoolmaster here in Hümmel used to take his pupils for a walk into the forest to marvel at this miraculous thing. Back then, they still had to be operated by two people but now the job has vastly sped up. The next major upheaval in Germany came with the winter storms of 1990. So many trees were knocked over that it was almost impossible to deal with them all. This was a job for a timber harvester. These mighty beasts, which had dominated commercial felling in Scandinavia since the 1980s, can grab hold of a tree, saw it off at the base, strip off the bark, cut the trunk into sections of the desired length, and stack them in neat piles. The work of twelve forest workers (even equipped with their state-of-the-art chainsaws) was replaced by one enormous robotic monstrosity.

With the storms of 1990, when millions of spruce trees fell to the ground, several harvesters were imported to cope with the huge volume of work. After the clean-up operation, they stood around unemployed for a while, but since they were cheaper to operate than human labour, the latter had to go. Since then, harvesters have dominated logging to an ever-increasing extent, while the number of forestry workers has continued to decline. And with them we have lost the romanticism of the lumberjack way of life. It is certainly amazing to see how a harvester tears through entire plantations at great speed, tossing and turning logs in their tongs as though they were as light as a feather. But there is something missing in the forest: no smoke from the workers' fire at lunchtime, no shouts of 'Timber!' warning of falling trees. Instead, there is only the monotonous hum of the harvester's engine, interrupted only by the brief whir of the built-in saw. And in the machine's wake lie desolate tracks, crushed under the machine's weight.

But that's just the damage we can see. Under the weight of the vehicle, the delicate soil is crushed to a depth of two metres. Pores are pressed together, air channels closed off. The impact for the underground micro-organisms is a grim death by suffocation. In addition to this, compressed soil can barely store any water, which is fatal for the trees in the following hot summers. Drought makes spruce and pine trees vulnerable to attack by parasites such as the dreaded bark beetle, which the tree would usually be able to fend off with its sap. When a tiny insect bores into the bark, it would normally be drowned in a sticky droplet of resin. But when the soil is dry, the tree is thirsty and has little moisture left for this line of defence. The beetle can nibble away merrily and summon its friends to the feast with its scent signal. Together they seal the fate of the tree within a few days. There is still lamentably little attention given to this consequence of using heavy machinery in the forest. When will the soil recover from this level of damage? According to geologists, full regeneration may not happen until after the next ice age, when frost and advancing glaciers thoroughly loosen the layers of earth.

Away from the landscape of industrial felling, though, the bucolic charm of the woodcutter lives on. There is little romance left in commercial logging, but an enthusiasm for doing it the good old way survives in the private sector, and it's worth remembering that almost half of Germany's woodlands are in the hands of private owners. The plots tend to be either inherited or bought for firewood. And because half a hectare (5,000 square metres) is enough to heat a modern detached house, it's little wonder that there are now about two million proud silviculturalists in Germany – people engaged in the cultivation of woodlands. Across the country, amateur woodcutters are heading into the woods with the families at the weekends to toil the day away, stopping only for a hearty lunch. This resurgence of small-scale forestry only dates back a few years and yet it has already found its way onto our screens. To my surprise, I suddenly started seeing TV adverts for the kind of chainsaws which had previously been considered niche equipment for commercial use. Hundreds of thousands of amateur foresters were signing up for courses at the state forestry offices in the hope of obtaining a chainsaw licence. Although this is not required for domestic use, if you don't own a forest plot of your own, you can often purchase a so-called 'woodlot' in a public forest, where you would need a licence to use a chainsaw. These are designated plots of land where trees are felled but left for lot holders to strip off the branches, saw into logs and transport using their own muscle power and wheelbarrows. A log fire warms you up twice: once when you're chopping the logs and again when you burn them.

It remains to be seen whether maintaining a wood in this way is economically beneficial. People are often tempted to buy equipment that vastly exceeds the requirements for the task at hand (and it's usually the men who get carried away; the number of women involved in amateur forestry is still very small). The chainsaws they choose are usually much too powerful and heavy. Instead of opting for a small, amateur model, which is comfortable to hold and perfectly adequate for slender trunks and branches, hobby foresters can't resist the lure of high-horsepower professional machines. If you spend all day working with a whopper of a power tool, you're going to ache by the end of the day. Never mind, they seem to reason, if you're going to do it, you may as well go the whole hog. And are they happy to drive into the woods with an SUV and a trailer? Are you kidding? Your regular family car isn't rustic enough for today's lumberjack; he has to have his own tractor for the job. With a built-in winch and a huge log splitter to chop the trunks into stackable logs. The cost of all this equipment of course vastly exceeds the cost of several years' worth of pre-prepared firewood.

And yet, I can understand anyone who does it well. It is the work, the self-generated heating, which makes every log fire in the evening that much more idyllic. When you have hand-crafted each and every log yourself, don't you hear the sounds of the forest in the flickering fire? And wood is cheap! Even while oil prices are low, in Germany wood is still significantly cheaper. For example, in rural areas a cubic metre of beech wood including delivery costs less than fifty euros, which is how much you'd spend on oil to achieve the same calorific value. The equivalent would be 200 litres of heating oil, which tends to be around twenty-five euro cents per litre. Even the best-quality timber, which is too good for burning, costs less than that. The boom in firewood consumption does not seem to have resulted in a price increase, but only in an increase in imports. And since this imported wood is often linked to overexploitation, or at least comes from less sustainable sources, it comes at rock-bottom prices. There's not much you can do to check the source when it's come from abroad. At least if you buy your wood from your local forester, you can verify the forest is certified to the FSC standard. This is a kind of eco-label, where the requirements are slightly above the statutory legal requirements.

Another option is to have your firewood delivered to your home in the form of whole tree trunks. However, dealers always deliver by the truckload, which is about forty cubic metres. Once these logs are sawn, chopped and stacked, you end up with a wooden wall one metre wide, two metres high and twenty metres in length. A bit too much for your garden? Well, to heat a modern family house exclusively with wood, you'd need about ten cubic metres per year. Since the wood needs to dry out for at least two, or even better three, years before going on the fire, your minimum reserve swells to thirty cubic metres. Seen in this light, a truckload is just the thing. Then you'd have no need for a tractor or a trailer, but a chainsaw would still come in handy. At home in your garden, you wouldn't need a chainsaw licence, but I would recommend getting one anyway. Chainsaws are extremely dangerous, and the sharp teeth can cause a serious injury in a fraction of a second. Other injuries are caused by falling branches or trunks splitting apart. Even among professionals, the accident rate is so high that every year, one in three regular chainsaw users incurs an injury serious enough to warrant medical attention.

Despite all the warnings, I must say it is very satisfying to make your firewood yourself. It's not just the sawing. If the trunks are cut to logs of a metre in length, they also need to be split crossways to dry better. It helps to know what you're doing, because wood splits much more easily if you place the log the same way up as when it was growing, rather than upside down. The old woodcutter's wisdom says '*Der Holz reisst, wie der Vogel scheisst*' ('The wood splits, like the bird shits') – that is, from top to bottom. With time, you start to notice small cracks that you can get your axe into, reducing the effort you need to exert. You don't need a gym when you have log splitting as your regular exercise.

But of course, there's also the question of convenience. You can buy seasoned firewood from wholesalers or DIY stores that is already cut to size and sold in manageable quantities measured by the cubic metre. It can be delivered in boxes or crates, from which you can fill your basket for the fire or wood burner. Of course, buying your logs pre-seasoned costs considerably more, not least because there are hidden price hikes along the way. And we are faced with the crucial question: what is a cubic metre, anyway? While heating oil can be measured accurately with standard gauge measurements, when we buy firewood quantities, part of what we buy is the air in between the logs. A solid cubic metre (known as a *Festmeter* in the industry in Germany) of wood is a cube of pure wood where each dimension is one metre. This is the unit of measure used to quantify the volume of every felled trunk that is sold, so that the buyer knows how much raw material he is getting. So far this all seems rather obvious, but when the cubic metre measurement is used in retail it can be confusing, which is why I emphasise the difference. At this point we enter the realm of pure speculation. As consumers, when we buy a cubic metre, it is of roughly stacked wood. The logs are generally thinner, often already split and usually sawn into sections one to two metres long, making it too time-consuming and costly to accurately measure each piece. Instead, they just go ahead and measure the whole stack, but of course as there are gaps between the logs, a cubic metre ends up containing a lot of air. How big the gaps are depends on whether the logs are crooked or straight, and whether branch stumps have been sawn off to allow the logs to be stacked neatly on top of each other. On average, one cubic metre contains 30 per cent air, leaving you with only 70 per cent wood. Therefore, if you're comparing prices for a solid cubic metre and a chopped and stacked cubic metre, you'll need to convert accordingly.

In the last few years, a new pricing model has been added: the loose cubic metre. While logs are normally stacked for measurement, firewood is sold more and more often in a format that is seasoned and ready for the fire or burner, chopped to pieces thirty centimetres long and poured loose into lattice boxes or nets. This creates even more gaps, so that a loose cubic metre now contains at least 50 per cent air. And the consumers? I think it's very hard for them to judge. It is of course convenient to have the wood delivered ready to use.

Unfortunately, the overwhelming price advantage of firewood over heating oil and gas is lost in the processing. By way of example, a solid metre of beech stacked in the forest, ready to be driven away by the purchaser, costs around fifty-five euros. If this trunk is chopped to metre-long logs and stacked for the buyer to collect, it is now costs eighty euros (still per solid cubic metre). Cut to size and seasoned for the fire, and poured into a crate or net, it is now well over a hundred euros. Add to that the delivery fee of around ten euros per loose cubic metre, and the price has doubled. Of course, what you're paying for is all the work carried out along the way – a lot more than when you buy an unprepared tree trunk in the forest, and it's you that does the rest of the work. But why not join forces with the neighbours and get a truck delivery together? That way you share the costs and the work of preparing the wood, making it both more fun and more worthwhile financially.

Of course, there are even cheaper options. I have seen firewood sold at less than sixty euros per loose cubic metre. Such low prices tend to point to imports from dubious sources. Knowing how badly Romanian or Russian forests are being overexploited and how low the wages are for the processing, it's best avoided. Are you going to enjoy a cosy log fire knowing it's contributed to the destruction of ancient woodlands? The romance fades very quickly.

It's another kind of disappointment entirely when you buy firewood by the bag at the hardware store or garden centre. In very small quantities, the fuel price has rocketed, and besides, despite what they insist on the label, very often the wood is not sufficiently seasoned. If it hasn't been left long enough to dry, the logs are still so damp that mould forms even there on the pallet. Too much moisture in the wood is also the main cause of smoky chimneys; the smoke is not only a nuisance for the neighbours, it's also against the law. Wood should not contain more than 25 per cent moisture before burning, but even then it gives off three times the quantity of harmful emissions than if it were dried to less than 15 per cent. This can be done either artificially in a drying chamber (which also consumes energy) or simply by storing them in an airy place for two years – and here we have the same problem of space.

In any case, I enjoy preparing my own firewood. I get full logs delivered which I then split and stack to dry for two years, when I then saw them into pieces twenty-five centimetres long. I then stack a six-month supply at a time in my wood store, so that I can always quickly fill the basket to bring into the house even when I'm in a rush. Along the way, from start to finish when the wood goes on the fire, I will have held each log in my hands five times. You get to know the wood. I recognise difficult, knotty features that annoyed me when I was splitting it, but later when they go on the fire I see them as old friends who gave me no trouble at all. Who develops such an intimate relationship with their gas heating?

14. Conservation: Love with an Impact

RECENTLY I WAS WALKING WITH MY WIFE AND FRIENDS in the Harz National Park. After a half-mile stretch through the forest, we came to a clearing that turned out to be a large meadow. Before I could even look around, I noticed a triangular sign with a green border declaring this to be a 'Conservation Area'. Below was a sign explaining why the meadow was granted protected status: rare mountain flowers grew here and living in harmony with them were a large number of endangered animal species. They had been rescued with EU assistance and how that worked was clear to see: freshly cut tree stumps were drying in the sun and beside them, at the edge of the path, lay the remains of the crowns hacked into small pieces. I was annoyed – again. It's one thing to protect meadow flowers, but it's another to cut down forest trees for the sake of them. I wouldn't call that conservation.

By nature, Germany is a pure woodland, after all. Its many tree-free landscapes have been created almost exclusively by human activity. The forests were cleared centuries ago and the soil was so intensively used for agriculture that it was completely exhausted. There was no artificial fertiliser back then and a little manure was not enough for the barren fields. Over time, the only plants that could survive on this depleted soil were ones that could cope with a meagre nutrient supply. Delicate flowering plants migrated to us from south-eastern Europe, spreading in the wake of deforestation and soil depletion. But when artificial fertiliser was developed, it spelled the end for meadow flowers. The soil could be beefed up again and meadows, pastures and heathland were again brought under the plough. Now we're seeing nostalgia for the good old days in the form of meadows set aside with the designation of protected areas. Whether it's Lüneburg Heath, the alpine meadows of the Allgäu or this nature reserve I described in the Harz, what they all have in common is the continued struggle against the return of the forest.

How is it that such a forest-friendly nation has all these protected areas while the campaign against the trees continues apace? I think our nation shares a deep love of nature, but it's gone somewhat astray, distracted by the desire to save as many species as possible. The keyword is biodiversity. It is to be preserved at all costs, and whenever a species is threatened, rescue programmes pop up, such as in the meadows at Harz National Park. In fact, you could go up and down the country trying to protect rarities such as gentians, orchids or daffodils that are lost to modern agriculture. But let me clear up a major misunderstanding: biodiversity in itself has nothing to do with nature conservation. You can find the greatest biodiversity in the smallest space inside a zoo. It wouldn't occur to even the most active environmentalists to designate a zoo as a nature reserve and yet that is exactly what we call our national parks.

As soon as the capercaillie, or wood grouse, threatens to disappear from the Black Forest, for example, forests are cleared and a kind of artificial taiga is created to allow it to thrive again, along with dwarf shrubs such as bilberries. Besides insects, these berries are an important food source for grouse and their chicks. Yet warm-climate insects such as forest ants and also bilberry bushes were only able to spread into our part of Europe because of the large-scale deforestation in the Middle Ages. The dark forest floor under the shade of mighty beech crowns became sun-drenched biotopes where all sorts of herbs and shrubs thrived. So, where humans cleared the woodlands, the capercaillie followed. Today, the forest canopy is closing over again, as there is at least a partial return of the original deciduous trees. Unfortunately, this means the end for the capercaillie in many places, although it is by no means endangered globally. During one of our visits to Swedish Lapland, my family and I saw convincing evidence that the forests are still pretty well stocked with wood grouse (which, incidentally, are still eaten there, but that's another topic). In Central Europe, the capercaillie is only at home where there are taiga-like ecosystems; this is the case in a few high-altitude regions of the Alps just below the tree line. There, where the weather tends to be rather harsh, the proud capercaillie thrives to this day, but even outside its natural habitat it is a bird that people have a soft spot for. The Black Forest pays dearly for this affection for the wild grouse, because assisting this species means having to dispense with others. But which? Well, no one knows precisely, because there is

still much research to be done into the true forest species
of Central Europe. While hundreds of species of moss
mites, springtails and bristle worms are awaiting investigation, their populations are devastated whenever
thinning brings more light to the forest floor or when
the tree species composition changes. When unappetising, sour spruce needles trickle down instead of mild
beech leaves, they lose their appetite and they starve to
death. But who mourns the loss of minuscule mites?
These poor little critters, which only evoke associations
with dust allergies, are not fluffy or cute enough to
attract public research funding. And yet they should
perhaps be considered as a kind of plankton of the soil,
the start of the food chain that is indispensable for life
in the forest.

Although it may be well intentioned, to unleash a
major transformation of the forest for the sake of
protecting a single bird species brings with it the risk
of a local extinction of another species. This is the reason
for my personal criticism of the approach. It is reason-
able that laypeople might not have the full picture; they
rely on the judgement of experts, after all. Professionals,
on the other hand, should not be guided by their
sympathy for charming birds or colourful flowers, but
by their mission to preserve native ecosystems. We bear
the responsibility for protecting the relatively small area
of ancient beech forest – a global responsibility which
we still do not take anything like as seriously as we
should. Maybe this responsibility will finally be taken
seriously in the new national parks that have mush-
roomed in the last twenty-five years?

I was delighted when the Eifel National Park was established. For a country of its size and wealth, Germany has relatively few protected areas, so we can hardly criticise the treatment of the Amazon rainforest. Our responsibility is to preserve or restore the original beech forests, which, on a global scale, are minute. Germany was originally at the centre of their geographic range, and I say 'was' because regrettably there is no longer a single square metre of that original forest. Still, there remain a few old beech woods, where timber is harvested but which are otherwise very natural. Some of these woods form the core of new national parks such as Eifel National Park. But as these surviving woods are far too tiny to add up to the minimum area of one hundred square kilometres necessary to earn international recognition as a national park, nearby spruce plantations are included in the designated area of protection. That can be quite useful – they are also valuable trees, after all, if not the ones we're seeking to protect. The value of spruce is as protectors of beech saplings, as beech trees need to grow beneath a shady canopy for the first 100 to 200 years. In the forest I manage (not part of Eifel National Park) we also have spruce acting as step-parents to the deciduous young. We have had infestations of bark beetles, and as it is really too warm here for spruce and pine, if I didn't intervene, the entire conifer population would fall victim, leaving the beech saplings exposed. To prevent the spread of the beetles, we fell the infested trees and remove the bark to deprive the insects of their breeding ground. This preserves the old spruce trees as shade for the young beeches.

In a national park such an intervention would be counterproductive. Because even if were right to assume that in the long run a beech forest would emerge in our local climate, we should leave nature to find its own way there (or in another direction?). That's the exciting thing about it: seeing whether our theory was correct or if things turn out differently. And yet, things are already taking a different course, but not because nature is left to its own devices. Even in a national park, a designated nature reserve, it seems foresters can't resist meddling. And guess what? The rangers are doing what has long been banned in commercial forests: large-scale clear cutting. We see the same damage that we've seen everywhere else: harvesters crushing the soil, stripping the conifer trunks of their branches and dispatching the cut logs to the nearest sawmill. Does that sound like conservation to you?

As soon as the protection status for the park comes into force, the foresters in charge can't get rid of the conifers quickly enough. Suddenly they are declared non-native undesirables that need to be removed pronto. The ridiculous thing is that the sought-after deciduous forest relies on shade to grow; unlike spruce, which can happily get established on desolate clear-cut ground, germinating in their millions and quickly growing into another a coniferous forest, the precise thing the foresters are trying to avoid. The emergence of the primeval beech forests is postponed at least a hundred years into the future, because the saplings won't get established out in the open. Although it is never admitted publicly or written in any prospectus, the clear cutting is justified behind the scenes on the grounds that the conifers were felled to 'prevent the spread of bark beetles and fungus'. How convenient that they can then be sold on to the timber industry.

And if we didn't intervene? As we witness in the so-called 'core zones' (the untouched zones) of the Bavarian Forest National Park, the mass propagation of bark beetles leads to the death of spruce trees over a very large area. The dead trunks provide some shade at least, not really enough for beech saplings, but just enough for a mixture of deciduous and coniferous trees. Their chances of survival are now improved for another reason: the tangle of fallen trees makes it harder for deer to enter, leaving the beech saplings to grow up unmolested. The slowly rotting spruces also provide new humus, which stores water and helps the young trees to survive a dry summer. It will still take up to 500 years before the emergence of a primeval deciduous forest. But the forest is patient: for a tree, 500 years is just one generation.

If we want to consider the conservation of forests and woodlands in the round, then all that's missing is the animals. Not the wild animals, because we have already discussed them in detail. I'm thinking of our pets. They competed for millennia with their wild ancestors, although now it is the latter that are hopelessly outnumbered. Wolves, for example, are still rare in Germany, but this is true only of the wild variant. By contrast, their domesticated cousins – dogs – have a population of over ten million in Germany alone, according to the Federal Statistical Office. They also like to enjoy the forest. What is better for our four-legged friends than being let off the leash? But then what? Most dogs have a hunting instinct that comes from the past. With the exception perhaps of lap dogs, almost every breed was bred for a specific purpose, and most were linked to hunting. An impressive range of skills were catered for by domesticated gun dog breeds, including pointers, setters, retrievers and spaniels, known as flushing dogs: they are trained to flush game such as birds, that is drive them out of their hiding places for the hunter to shoot.

Although most dogs today are viewed more as family members than as helpers in the forest, every now and then a run among the trees stirs up their desire to chase rabbits or deer. If it's a healthy animal being pursued, a dog won't be able catch it. Deer, for example, don't head off in a straight line but loop round in a kind of circle. Their tracks cross over with older tracks they have previously left, so the dogs get confused and give up when they lose the trail. It only becomes dangerous for the forest dwellers if two or more dogs pick up the trail, as then they can corner their prey and come in for the kill. Unfortunately, dogs don't attack like wolves do and aim for the neck – that would lead to a mercifully quick death. No, dogs tend to sink their teeth into the rump or the side of their victim, resulting in severe injuries and a slow, painful death that might take days. In some states within Germany, it is compulsory to keep a dog on a lead, but even where it isn't and you can let your dog off the leash, it's important that it stays close enough for you to summon it back with a call or a whistle.

The leash has its benefits but also its downsides. I discovered this once when I came across an old collar and leash hanging from a young spruce tree. From the state of decay, it was clear they had been hanging there for years, the final relics of a tragic story. It would appear a dog had torn its leash out of its owner's hand and had got stuck on a spruce branch, out of earshot. There, the dog would have suffered a miserable death of starvation and finally been eaten by foxes or wild boars. So if your dog is allowed off the leash, and the local laws permit it, then best to do it properly: no collar and no leash.

15. The Woods in Bad Weather

WHAT DO YOU DO if you find yourself in a forest during a thunderstorm? Of course, it's advisable not to go out walking at all if the forecast predicts a storm, but if you are caught in one by surprise, what should you do? An old English folk saying warns, 'Beware of an oak, it draws the stroke; avoid an ash, it counts the flash; creep under the thorn, it can save you from harm,' while in German we have a proverb that says, '*Eichen sollst du weichen, Buchen sollst du suchen*' ('Keep away from oak, seek out beech'). This seems to stem from our ancestors' interpretation of lightning damage, which often affected oak trees, but never beeches. Does this mean beech trees are lightning-proof and safe to shelter under?

This folk wisdom is an illusion, because beeches are by no means immune to a lightning strike. Their smooth bark allows water to stream down the trunk in a continuous flow down towards the roots. In a heavy downpour, so much water can run down the trunk that a white foam forms at the base of the trunk. The bark of oak trees, on the other hand, is rough and cracked. As the rain runs down the tree, the flow of rainwater is split by the bumps on the bark into hundreds of tiny rivulets which are constantly interrupted. Lightning always seeks out the route with the best electrical conductivity, which is not the water running down the outside. Immediately under the bark is the tree's water transport system, the pipelines which pump water from the roots up to the crown, visible in the cross section of a tree trunk as the outer growth ring. The lightning makes its way into this conductive route, but the minuscule pipes cannot withstand this electrical charge and burst. The explosive force of the lightning bolt can be so violent that splinters of wood fly through the air like knives and get imbedded in neighbouring trees. Lightning scars can still be seen on the bark of an oak tree many years after it was struck and this was once taken as an indication that this species acted as a kind of lightning magnet. In fact, the probability of being hit by lightning is the same for all tree species, with the only decisive factor being height. So, avoid hilltops and don't seek shelter under trees that are particularly large and whose crown protrudes above the canopy of the forest.

More common than a thunderstorm is a normal rain shower. What do you do when a heavy downpour appears out of nowhere and you have neither waterproofs nor an umbrella? Then you're faced with a choice: which is the best tree to seek shelter under? In contrast to popular beliefs about lightning, there are in fact significant differences between various tree species. The branches of deciduous trees grow diagonally upwards so that the water runs down them and is led down the trunk to its own roots. Oak trees and beech trees are excellent water collectors, and that's why they're particularly soggy to stand beneath when it rains.

Besides, long after the sun comes out, water drops fall from the leaves, the origin of another German saying, '*Unter Laubbäumen regnet es immer zweimal*' ('Under broadleaf trees it always rains twice').

It's a different story with conifers. They hail originally from the far north, from regions where they are never short of moisture and where they have little need to collect rainwater with their branches. Instead what is important is stability in heavy snowfall. A heavy burden of snow could cause the entire crown to snap off. The branches of a conifer therefore grow horizontally and droop down at the end. When a lot of snow falls, the tree simply drops its 'arms' to its sides under the increasing weight, thus significantly reducing its footprint as seen from above. And in the rain? Most of the water runs out along the branches and away from the trunk. This means it is always particularly dry underneath a conifer – something you can take advantage of in a shower. The closer you are to the trunk of a spruce, the drier you will be. For the tree, however, this is a drawback in our more southerly latitudes, because it 'gives away' precious water. This, combined with the lack of water stored in the compressed soil, means that conifers are quick to fall victim to drought.

If you are in a deciduous forest, you can listen out for the weather forecast as announced by the birds. The chaffinch has a range of calls in its repertoire. Its fair-weather call is a cheerful melody, but as soon as rain is on its way, this chirpy sequence switches to an ugly-sounding '*Raaatch*'.

Quiet, foggy weather can be deceptive. You're wandering around under old, mighty trees – what's the worst that could happen? It makes for a very mystical, fairy-tale atmosphere, when dark tree trunks appear before you, and sometimes disappear again, amidst slowly rising wisps of steam. The haze also dampens all sound, making you feel completely alone with nature. But in very dense fog you occasionally hear a loud crash, or a thud, as though something big has hit the forest floor. And that is precisely what has happened. It is branches as thick as your arm falling from the crowns of mighty deciduous trees. In calm weather? Nobody would bat an eyelid if there was a storm blowing, but you don't expect such deadly danger to fall from above in calm conditions. What causes it is the high humidity. Dead, rotten branches absorb the water droplets like a sponge. Their stability is greatly reduced by fungi, bacteria and beetle larvae, as the small organisms tirelessly nibble away at the wood, leaving behind a soft, squidgy sediment. The absorbed moisture is like the straw that breaks the camel's back. The extra weight overstretches the weakened wood that remains, it breaks, and the branch crashes to the ground.

Ice crystals on branches, or rather hoarfrost, is another deceptive phenomenon. As lovely as it looks, it can be fatally dangerous – not only for you, but also for the trees. Here, fog comes into play again, but this time at temperatures below freezing. If such weather conditions continue for days, more and more ice crystals stick to the branches. This continues until even thick branches break off, or, in some cases, the whole tree cracks apart. This sort of damage occurs every five to ten years. However, I've only seen hoarfrost once in my career so far. It had rained for three days – a light, harmless steady sort of drizzle at minus-3°C – and the whole forest was glazed with a centimetre-thick crust of ice. Younger trees in particular were bent down to the ground under the weight, while the tops of some of the older conifers had snapped off and fallen. There was no one out walking because the paths had turned into lethal ice sheets like mirrors.

So, what are the relative risks? On a frosty day, the hazards of walking in the woods are limited and it's unlikely to be more hazardous than the risk of being struck by lightning. Is fog more dangerous than a thunderstorm? Definitely not. I would go out for a walk without hesitation in fog but, in a thunderstorm, you should definitely stay at home because of the risk of whole trees breaking apart and being overturned.

16. Litterbugs and Forest Fires

GLASS BOTTLES ARE ALWAYS GETTING PEOPLE WORKED UP in Germany. Why? And what have they got to do with the forest? Some (perhaps overly cautious) people argue that discarded glass bottles are responsible for forest fires. The thick base of the bottle acts like a magnifying glass, they say, and it concentrates the sun's rays, heating up the focal point to such a degree that the dry plant litter could ignite, like tinder. That sounds logical; I certainly used to experiment with starting a fire with a magnifying glass as a young lad and remember the thrill every time the little white dot of focussed sunlight burned glowing holes in the newspaper. If a thick bottle base is curved in the right way, it could act in a similar way, in theory at least. But I've never heard of a forest fire being started by a discarded bottle. The hypothesis was also once investigated by the newspaper *Die Zeit*, which commissioned an experiment carried out by the German Weather Service. They shone light through glass bottle bottoms, selected for having the right shape, trying to heat a fuel source to 200°C, the minimum temperature for autoignition, also known as the kindling point. Despite various attempts, they only succeeded in heating the fuel to 80°C, and that was under the best laboratory conditions.[25]

So we can rule out glass bottles as fire starters, and yet there are plenty of other reasons why you should never dispose of them in the wild. I'm always finding old fly-tipping sites from the 1950s and 1960s. At that time, villagers simply dumped their waste down the nearest slope – out of sight, out of mind. These wild dumps were covered over with soil after the introduction of regulated waste disposal, and the sins of the past were buried and forgotten. I have long wondered how nature-loving people could do such a thing. Well, I suppose the reason is that until the widespread use of plastic, most of the waste we threw away was compostable. Wood, leather, paper – it would all decompose eventually. In those days, then, it wasn't quite so bad to dump everything in a heap in the wild. Valuable glass and metal were collected to be reused or recycled, so it wasn't left lying around to blight the landscape.

But in the early post-war decades, with rising prosperity, we shifted towards a disposable mentality, while persisting with the old tradition of fly-tipping behind the houses. This era has bestowed us with lasting glass and metal finds, where the shallow layer of soil has been washed away in many places. If you're not looking where you're going, before you know it you've got a shard in your foot. We humans wear shoes, at least; wild animals don't have this protection. For their sake above all, we should take care not to bring any more litter into the forest, even if isn't necessarily a fire hazard.

One other comment about forest fires: they are a natural feature of a hot, dry summer. To understand them, there's a little experiment you can do at home. Try setting fire to a green beech or oak branch. Go ahead and try! You'll quickly find you can't. Living deciduous trees don't burn, no matter how hard you try. And because even lightning can't set a living tree ablaze, the broadleaf forests of Central Europe are not naturally prone to forest fires.

From time to time we hear about 'forest fires' in the news, even if it's not a genuine forest that has flared up. As we've seen, plantations of spruce, pine and other conifers account for more than half of Germany's 'forests' – monotonous rows of planted trees that have little natural about them. The needles, bark and wood all contain essential oils and resins that burn easily. Besides this, at the feet of the trunks there accumulates a thick layer of dry twigs, bark and needles, because our domestic soil organisms can't properly digest this bitter plant matter. This dry layer of tree litter is the perfect tinder just waiting for a discarded cigarette butt. In fact, in the heat of summer, the proverbial spark is enough to have hundreds of trees up in flames. The fact that it will rarely spread further is due to vigilant fire surveillance, with every wisp of smoke swiftly reported to the nearest fire department.

Is there such thing as 'acceptable' litter? Is there anything that we can safely dispose of in the forest? This is a question people always ask themselves when they're out hiking. After all, who wants to put a damp banana skin or apple core back in their backpack? Or a used paper tissue, soggy and already decomposing? Surely no one would notice if you quickly tossed it into the under-growth? It is organic matter after all and within a few months it will return to humus. Well, I would advise against such a move for several reasons. Fruit peels are coated with spray or wax, to make the surface shiny and attractive. These additives make it hard for the peel to decompose and leave chemical compounds in the soil which were not previously present.

The same is true of used tissues, but these also have another undesirable side effect: being bright white, they stand out and signal to everyone that there's litter here. And litter attracts more litter. This is the reason why it's rare to find a bin in a forest these days, although once they were ubiquitous. When the bin is full, people start leaving rubbish on the ground next to it. Whereas, if there is no bin in the first place, walkers take their litter home with them – unless previous visitors have left theirs on the ground in a corner, in which case people reason that it won't make much difference if they add a bit more, right? So, basically I would urge walkers to put any form of waste – organic or not – right back where it came from: in your own backpack.

17. Lost Without a Watch or a Compass?

I HAVE TO SAY, I feel lost without my watch. Until it's there on my wrist, I don't feel properly dressed in the morning. But it's not just me worrying I'll be late for my appointments (yes, even a forester has a schedule to work to). I also love the ticking of mechanical clocks on the wall and the mantelpiece; the muffled chime on the hour is an echo of times gone by.

Whether it's an analogue antique or the digital display on your phone – we have all internalised the clock's routine. But our understanding of time is flawed in several ways, which can confuse matters if you're in the woods without a watch and want to work out the time from the position of the sun. The first problem is astronomical in nature. The time we read on our watches here in Germany is Central European Time (CET), which corresponds to the position of the sun at 15° east of Greenwich. For example, when it is 12 o'clock, the sun is precisely due south everywhere on this line of longitude. In Germany, this would be Görlitz on the German–Polish border and in Austria, it's Gmünd in the Waldviertel (forest quarter), whereas Switzerland doesn't feature at all on this line of longitude. This means that for everyone else using CET, which is most of Europe, the sun is not due south at 12 noon. Here in Hümmel, which is in the far west of Germany, so on the other side of the country from Görlitz, we lag behind CET by about half an hour. That means that when it's 12:00 CET, the true local time according to the position of the sun is only 11:30 a.m. The earth still needs to turn for another half an hour before the sun is exactly at its highest point, due south: the sun reaches its zenith for us at 12.30.

This time lag is exacerbated when we come to the second built-in error, when we shift the clocks forward for Summer Time. The clock time is shifted one hour later, putting it out of kilter with the solar time by sixty minutes. So now if I look at the sky at noon, the sun is in fact only at the position where it ought to be at 10:30 a.m. Why am I talking about all of this? Because the forest doesn't recognise human clocks, but of course aligns its rhythm with the true position of the sun. Just like us, the forest knows the difference between day and night, between dawn and dusk, and all the gradations in between. We only need to tune in to the bird clock to see how precisely birds can distinguish increasing levels of brightness, as they melodically chart the sun's progress through the sky. So that every songbird has a chance of being heard, each species has its own time slot – or rather its own slot depending on the sun's elevation – when it really gets going in the morning. While the skylark starts an hour and a half before sunrise, it is another sixty minutes before the chiffchaff gets into the mood. If you can identify the species in your local woods by their song, you can put together a very personal bird clock based on their dawn chorus. One thing, however, is common to all birds: when the sun rises over the horizon, they all join in. This natural musical clock is only suitable for early risers and, of course, only works in the summer months.

So, we've dealt with time when you don't have a watch. What about orienteering by the tools of nature when you don't have a compass to hand? Do you remember the old Scout rule about using moss as a compass? Supposedly when moss grows on a tree trunk, this is the side that the weather comes from. And since the rain mostly comes on the west wind where we are, tree trunks get especially damp on that side. Moss loves moisture and therefore grows on the west, offering you the perfect natural compass.

This is true in many cases, but if you rely on it in the forest and orientate yourself exclusively by moss growth, you are guaranteed to get lost. Because under the protective forest canopy, there is no wind and the rain usually falls vertically. The position of moss on forest trees is therefore determined by a very different set of circumstances. Trees rarely grow exactly straight up; most tree trunks have a very slight, banana-shaped curve to them. As we've already mentioned, deciduous trees collect the water in their branches and direct it down the trunk to the roots. The water flow is influenced by the curvature of the trunk, so that there'll be a small stream running along the upper side, while the water droplets simply fall off the underside, leaving the bark on this side with very little moisture. That means no moss will grow there, while a thick cushion of moss will form on the topside. All this indicates is the curve of the trunk, nothing about the points of the compass. And because each tree bends differently, the moss will point you this way ... that way ... until you're thoroughly lost. Try it yourself next time you're out for a walk in the woods. Incidentally, moss is less likely to grow on coniferous trees because their branches direct the rain away from the trunk. At least they don't add to the confusion of Scouts out orienteering.

Realistically, though, you can't get lost in the forests of Central Europe. If you have a mobile phone or a GPS device with you, your route is tracked, so you can check your location – as long as the battery doesn't run out! It might be technology from the last millennium, but a good old hiking map is usually more reliable. You might take the wrong fork in a path somewhere, but to get so lost that you spend days wandering in the woods alone is not, statistically speaking, ever going to happen. Though we may call them forests, most of our forested spaces are actually tiny woods. Take a look at the patch-work rug landscape in an aerial photo on the internet and you'll see how small these islands of green really are. Experts say it's only when you're a kilometre away from the nearest meadow, road or settlement that you get a genuine forest climate without any outside inter-ference. One kilometre? In most cases you're out of the forest again on the other side.

Another yardstick for measuring our woodlands is the territories of wild animals. When we consider that even the small wildcat, which feeds mainly on mice, needs a territory of five to ten square kilometres, then we get a sense of the scale of a genuine forest: it should be several times bigger than the territory of one wildcat. And even then, that's not enough. According to Germany's National Forest Inventory, there are about thirteen kilometres of surfaced tracks per square kilometre of forest. They cover every last nook and cranny, so that trucks can carry away the felled wood all year round. Branching off from these tracks are further aisles cut through woods for the enormous harvesters. Because these 'backstreets' occur at intervals of twenty metres, they add up to an incredible fifty kilometres of track per square kilometre. A wilderness without paths and tracks looks quite different. The worst thing that can happen to walkers in Germany is that they emerge from the woods in the wrong village and have to order a taxi to get back to where they left their car. If, despite everything, you're unsure of your location but want to avoid going around in circles, follow this simple rule of thumb: keep going downhill until you're back on a solid road. You may find yourself taking a few detours, but at least you won't be wandering aimlessly through the forest. If you come across a river or a stream en route, follow it in the direction of flow – that is, also downhill.

It's reassuring to know it's unlikely to happen, and yet the idea of getting lost for days on end, without any food or a mobile phone, has something strangely appealing about it. It's an intriguing hypothesis – what would happen if you really did get lost in the wild?

Would the forest provide enough sustenance for us if we needed to survive in the wild? And how long could we last? It can be fun to give it a try, even if in Europe you're unlikely to find yourself needing to put your forest survival skills to the test.

18. Forest Survival Skills

FOR A WHILE I USED TO RUN regular survival skills workshops. The only things participants were allowed to bring with them were a sleeping bag, a cup and a knife. We would walk to a remote area of the forest and spend the whole weekend there as a group. Since these events usually took place between May and September, you'd think the foraging prospects would be good. Mushrooms, berries, nuts – what more do you need to keep hunger at bay for forty-eight hours? Well, as tasty as this forest salad sounds, I wouldn't get your hopes up. These fruits of the forest are only available for a few weeks a year, and they're not particularly high in calories, with the exception of nuts. And good luck harvesting nuts; the squirrels get most of them before they're even ripe.

So, if we're going to last the weekend, we need to keep our eyes open for other sources of nutrition. The cambium of the spruce tree is suitably abundant. This is the growth layer of the tree, located just under the bark. This layer of cells produces wood on the inside and on the outside it forms the bark. In winter, the tree contains little water and the bark clings tightly to the trunk. From March onwards, however, as soon as the spruce trees come out of hibernation and they're pumping water out of the ground again, you can quite easily peel off the outer bark with the help of your knife. The height of this phase is in May, when you can peel whole strips off. To avoid damaging a living tree, you could test this method out on a spruce that has been overturned in the winter storms. Once you've peeled away the bark, you expose the shiny wood within. So, where do you find this nutritious cambium, then? It's right there: this shiny layer that glistens with juicy sap. Running the blade of your knife flat against the wood, you can scrape off milky strips – *voilà!* It tastes a bit like resinous carrots and contains not only vitamins but also sugar and other carbohydrates. Cambium is one of the most nutritious food sources in the forest, per gram consumed, and in terms of taste it is also a culinary highlight.

Resinous carrots aren't making your mouth water? Well, it is what it is, and if we don't fancy it we only have ourselves to blame. Grazing in nature means relying on food that can be bitter or sour in taste, tough and fibrous in texture, and only found in small quantities, so you have to spend most of the day foraging. Cambium is a blessing indeed. We no longer appreciate it because of how our food has evolved. Recent decades have seen merciless competition for our appetites. We are led by our palates in a search for calorific delicacies – this desire is our genetic heritage from the distant past. We instinctively crave foods that are fatty, sweet and salty, compressed carbohydrates. This urge would have made sense 10,000 years ago, after all, when a calorie bomb was a rare thing to find, and if you stumbled across one you had to eat it immediately. Faced with supermarkets crammed to the rafters with calories, there is no reason to binge eat in this way, but we struggle to switch off our pre-programmed instincts. Instead, our food has been improved and optimised to make it fit as closely as possible to our unconscious longing for taste and calories. Products only survive in the market when they tick these boxes, and even then they're only popular until something even tastier emerges. The impact on our diet is that everything we eat tastes more or less the same.

Is that a bit of an exaggeration? The proof is out there in the countryside. Try fresh rowan berries, a ripe sloe or a salad of daisies and dandelions. Even at the thought of it, I can't help but screw my mouth up in distaste; after all, civilisation has moulded my preferences too. Seen in this light, cambium is indeed a blessing from nature, at least between March and July. Later in the summer, the trees are already preparing for winter again and the sapwood starts to dry out again. By this time, it's hard to scrape off more than a few scraps of bark and there's little sign of any juicy cambium underneath.

We are omnivores, though, and there are other delicacies to be sampled when we look to the animal kingdom. What about longhorn beetle larvae? These are flat grubs, several centimetres long, white with a dark brown head. Their flat shape allows them to creep along beneath the bark of dead trees on the search for the last remaining nutrients. They shred the bark with their large teeth and clean out the dried cambium. These larvae are tiny protein bombs and if you're going to get by in the forest, you don't have much choice, so get stuck in! Except it's best not to down it in one; after all, this is a foodstuff that bites back. I'd recommend you bite the head off first, chew and swallow, and then enjoy the rest. The maggots have a nutty, earthy flavour, and once you've got the head out of the way, the long-horn beetle larva is right up there for taste, in the same league as the juicy cambium. The best place to find this nutritious grub is on any tree trunks or large logs left behind to decay after the last timber harvest. Roll the log over to reveal the side that was lying on the damp floor. With the help of a pocket knife, you can slice off whole chunks of soft, crumbly bark to reveal the pale larvae living inside.

If you have no luck with the beetle larvae, at least there are always woodlice in abundant quantities. You should probably try and banish all associations with your doormat or the steps down to the cellar, otherwise you'll hardly be in the mood for your snack. The fact that woodlice are related to crustaceans is immediately obvious from the taste, but only if they're consumed raw. To make the eating process a little less gruesome for the ingredients and a little more palatable for the diner, you could also lightly fry them in a pan with a little oil. I always have these two luxuries with me on my survival weekends to make the transition between civilisation and wilderness a little easier. And in fact, after a few seconds over the fire, these pan-fried insects taste like crisps. If you sprinkle on a little salt, and don't look too hard, you could easily forget what it is you're eating. But this takes us back to the beginning of this chapter and the evolution of our diet. Let's be honest with ourselves: who's going to find themselves fighting to survive in the wild armed with a frying pan and some oil and salt?

I've often heard people say, 'I can't eat that now, but if I had to, in an emergency, I would.' Actually, I think it's the other way around. On my tours, I always found that participants were bravest at trying things on the first afternoon. They were still full from their last meal at home or the rest stop on the way there, and every larval encounter was a humorous test of courage. However, as the second day passed by, when stomachs had been rumbling for a while and everyone was feeling weary from physical activity, the will to experiment was gone. Larvae? 'No, thank you,' they would say. 'The hunger is bearable and I'll be home again tomorrow.' Most prefer to lie down on the brushwood mattress and try to ignore their growling stomachs by taking a nap.

Incidentally, red forest anthills are also very productive. The ants scurry around in their tens of thousands and all you need to do is pick them up. You might want to finish your victims off first by squeezing them between your fingers; that way, they won't be able to bite your tongue. But do be careful when crouching down by a red forest ant nest. In no time at all you'll have them crawling over your shoes and up your trouser legs – on the inside. You can imagine how painful it is when they take a bite in the groin area.

And what about shooting or trapping larger animals or birds? Apart from the fact that, in Germany, you need a hunting licence as well as a permit for the area you are in, the chances of securing enough meat to survive are relatively low. It may take days before you get something in your sights and by then you would be exhausted. This raises the same question as the pan-fried larvae: who lugs a rifle around with them in the woods on the off-chance that they need to feed themselves in an emergency? No, sticking to the small woodland dwellers is much safer and more productive. And if you don't want to kill or eat any animals at all, insects or otherwise? Then you're not going to have a lot of options besides the cambium. Beechnuts are a delicacy when roasted (not raw!), but you'll only find them in autumn once every three to five years. Since the seeds contain about 50 per cent oil, you could get by quite well on them. Acorns are basically poisonous. But if you peel them and boil them several times, each time changing the water and flushing away the tannins, then you can eventually make these little calorie bombs safe to eat. They can even be dried and ground as a flour substitute. However, since oak trees also only produce fruit every three to five years, you would be very lucky if you found sufficient quantities.

The roots of wild herbs are perfectly delectable, such as the dandelion. These pale, thin storage organs need to be thoroughly washed, and even then you'll find you get an unavoidable crunch of soil in your teeth. If you cut the roots into thin slices, carefully roast them and then grind them to a powder, using your cup against the bottom of the pan, you can brew a kind of coffee substitute. It is brown, tastes bittersweet and might remind you a little of home, at least after a few days in the woods.

Mushrooms are not worth the bother without oil and butter, because their calorie content is practically zero. To be more precise, our digestive system struggles to unlock the calories, which is why they are excreted again only half-digested. And berries? Wouldn't it be a dream to stroll through the woods stuffing sweet black-berries and tiny wild strawberries into your mouth all day long? I once had a group that did precisely that during a survival skills weekend. It was a very hot day in July. We stumbled across a clearing overgrown with blackberries. On the bushes hung large, shiny black fruit – overripe and bursting with juice. Irresistible! We stopped and the group gorged themselves with berries, hoping to fill themselves up. But two hours later, most were hungry again and many were sick from an excess of fruit acids in the stomach.

Keeping yourself hydrated is a priority over eating. Water is the most important form of sustenance and we can't last longer than three days without a drink. Clearly, dying of thirst is rarely a risk in most of Europe, but this is a thought experiment that it's useful to consider when you're out and about in the wild. When I was hiking in the English Lake District years ago, we would have given anything to find some reliably clean spring water. We were each provided with a lovely packed lunch at our small bed-and-breakfast. We popped the bags into our backpacks without checking and it was only when we stopped for our first break up in the fells that we realised our mistake. We had plenty to eat but all we had to drink was one small carton of apple juice per person. We should have checked the contents of the packed lunch at breakfast and brought extra water. So of course we polished off the lunch pretty quickly, and for the rest of our long hike we were tormented by increasingly unbearable thirst. Not that we weren't surrounded by gurgling mountain streams. Every fifteen minutes we passed running water. But we were also surrounded by thousands of sheep and were well aware that their excrement ran off the fields into the streams. Such a shame. So we battled on and rushed into the first café we found in the valley, where we filled up on water and lemonade.

If you're lucky, you might come across a genuine river source on your hike. If the water trickles out of the earth among the trees and not in an excrement-filled pasture, then it should usually be clean enough to drink. Sometimes the name of the stream, if you can find it out, will help you identify whether water is unsafe to drink. In one of the Eifel region reserves where I was an apprentice, the area around one such brook was called *Em Dünndrisser*, which translates roughly as 'the place that gives you the runs'. Foresters clearly had some bad experiences after sipping this water centuries ago. But how often do you have the river and stream names to hand when you're walking through the woods? Besides, the water quality may well have changed since the days when it acquired its name. A more useful technique is to have a look underwater. The minuscule creatures that reside in the water also reveal something about the quality. Take stonefly larvae, for example. These insect nymphs live in streams for most of their lives (about a year) before emerging onto dry land to shed their skins and fly around for a few days. As adults, they mate, lay eggs and die. The nymphs slowly crawl about on the stream bed, especially among stones. If you find one of these flat, greyish brown insects with three pairs of legs and two long tail threads, you can lift it up on the stone it's sitting on and have a closer look. Seeing these larvae is a good sign that the water is clean enough to drink.

Salamanders are another good indicator of clean water. The small, delicate larvae look more like newts with their four legs and long tail, but – unlike newts – they have very faint dark spots on their skin and a pale yellow patch at the top of the legs. The young salamanders need a stream with very clean water, which is one of the reasons why they have become so rare. We regularly see adult salamanders near our house, especially in the night after a rain shower, when they go out on the prowl for snails and other small animals. Whenever we return late at night from visiting friends, we have to be careful not to accidentally trample on any on the driveway. If you have a regular visit from a salamander at home or come across them in certain places in the forest, you could keep a photo diary to chart their progress. Fire salamanders can be uniquely identified by their distinctive yellow–black colouration, and once you start to recognise your local salamanders you will be able to identify them by their spots and stripes even years later. And it can be several years, because a sala-mander lifespan can stretch over several decades (in captivity they've been known to live up to fifty years). It's therefore worth keeping photos to help you recognise old friends.

Even in summer, it gets cold when night falls over the forest and our thoughts turn to lighting a campfire, because feeling cold is at least as bad as being hungry. When we read about how easily forest fires spread, it's tempting to imagine we can get a campfire going without too much effort. Easier said than done. It takes some practice and even then it can often take a while to get a fire going, especially in wet weather. Everything is damp with cool water droplets and then you get a gentle breeze which keeps blowing out the lighter flame. Would you even use a lighter if you're honing your survival skills? You could, if you felt inclined, take a more archaic approach. There's more preparation involved, but you can get a fire going faster.

You'll need to assemble your fire-lighting kit at home, but you never know when it will come in handy. First, you need an old tin – the kind you get boiled sweets in is ideal. Pierce a hole in the lid with a nail. Stuff it full with a few small shredded scraps of cotton, e.g. from an old vest. Next time you have a barbecue, place the tin, with the lid on, at the edge of the smoul-dering embers and wait. You'll see it give off a tiny

white wisp of water vapour which will gradually disappear. Now put the tin aside to cool down, still with the
lid on. The fabric inside is now charred and is ready to
use as your char cloth tinder. Before you set off for your
hike, you'll also need to stock your kit with a piece of
flint, such as you might find on a pebbly beach. Finally,
you'll need a piece of carbon steel to strike against the
flint, ideally one that is forged into a handle shape which
you can easily grip. You can buy something suitable
online or look out for one at a car boot sale or salvage
yard. All you need now is some hemp rope (e.g. caulking
hemp from a hardware store), from which you can tease
out some threads to make it rougher and easier to ignite,
and your kit resembles something people might have
carried with them 2,000 years ago, as this uses the same
principles for lighting fires as they did back then.

To produce a spark, prepare a rough ball of hemp
and some kindling. Hold the piece of flint in one hand
with a swatch of char cloth tinder held against it with
your thumb. Hold your carbon steel fire striker in the
other hand. Now strike the steel against a sharp edge
of the flint, downwards with a stroking motion, which
should produce some sparks. The aim is to get a spark
to hit the char cloth, setting the fibres smouldering.
Now tuck this smouldering tinder inside the hemp and
blow gently and steadily until the smouldering spreads
and the hemp suddenly ignites. Pile kindling over it and
your campfire is good to go. With a little practice, this
ancient method of fire starting is very reliable and suits
non-emergency situations too, of course. It can be exciting

to kindle a fire in this way to bake campfire bread twists or grill sausages – always a highlight, especially for children.

The question on a rainy day is where do you find dry wood for kindling? The twigs and branches on the ground are soaked through and even if the hemp ball takes well, it will only keep burning for so long before it fizzles out. The answer grows on the trees, especially evergreen conifers such as spruce or pine. Their shady branch structure means that the trunks usually remain dry and around the base of the trunk you should find plenty of small twigs and dead branches. These will burn even after days of wet weather, and once you've got a fire going with their help, you can add moister branches later.

In principle, you should only light a fire on cleared ground. To do this, scrape away all the conifer needles, leaves and humus, and surround the fire circle with rocks if you can. You should take these precautions because while the upper layer of the forest floor will be drenched by rain and therefore safe, in some arid areas, it can be bone dry deeper down. Moss in particular tends to harbour easily flammable plant remains underneath. I once had to drive out into the woods in the winter, while it was raining, to extinguish the remnants of a fire, which some hikers had left smouldering after a rest stop. I had a twenty-gallon canister of water with me, and even that wasn't enough, as it quickly turned out. Although the blaze was barely half a metre in diameter, the smouldering continued to spread beneath the damp layer of moss, which acted like an impregnated roof blocking the water I was pouring over it and diverting it away. It was only by completely clearing the area of moss that I managed to put the fire out. However much we enjoy a campfire, this is why we need to always ensure that every last ember is put out before we set off.

So, everything has gone to plan: you've found enough to eat and you're feeling nice and full. Your digestion seems to be working just fine, which means that sooner or later we need to think about the next stage. But who wants to do their business without any toilet paper? The unfamiliar diet can also cause some problems in this regard, because – well, let's not beat about the bush – the stool is likely to be on the loose side. But here, too, nature is on hand to help. Look to the soft moss flannels that grow on old tree stumps. You can peel off a strip like a scrap of cloth and as it has a similar tensile strength to toilet paper, you can use it in precisely the same way. If it has recently rained or it's damp with the morning dew, then your moss always doubles up as a convenient natural wet wipe. As a friendly gesture towards other hikers, it would be thoughtful to bury the remains after use.

You should take care to choose your spot wisely when you need to do your business. And by that I don't just mean look out for wild game cameras installed by hunters who must get hours of fun out of the footage – no, it's the little pests you need to bear in mind. Squatting with your trousers down lends you a certain vulnerability to small mosquitoes and biting midges, which love to pounce on bare skin, especially if the forest in question is in a damp but slightly sunny valley. Shady areas are more suitable, ideally on or towards the top of a hill. Best of all is a spot with a good breeze – as mentioned above – as they can't stand the wind and they risk getting blown away.

So you've filled yourself up, responded to the call of nature, and that's pretty much the whole day spent attending to your basic needs. There are modest nibbles to be found, but some really are tiny, and to feel full you'd have to spend hours foraging for them. But as night looms there is still the question of where you're going to sleep. If you have an axe with you, you could fell a small conifer – a fir or a Douglas fir, say. Chop off the green branches and these can be used to construct an evergreen bough bed. The greenery should be laid on the ground so that curve faces upwards; this way the wood has a spring to it like a slatted frame bed. Lay the thick woody branches to the sides to form the outer edges of the 'bed', so that the soft green needles overlap to form your mattress. The thicker the stack, the more comfortable it will be.

But take care as you build your nest! Even after showing them how to do it, I've often seen survival skills course participants carelessly throwing branches over each other in a criss-cross formation. This is going to be painful at night, when the woody branches – the thickness of your thumb – press so hard against your back that you'll toss and turn like the princess on the pea. If, on the other hand, the branches are properly arranged to form a kind of bath tub shape, then you'll be comfortably embraced as you lay down in it. This is important, because somehow there is always a slight slope and the whole thing risks loosening every time you turn in the night, leaving you waking up in the morning on the bare forest floor. But if it's securely constructed it should hold together, keeping you cosy in the middle between the branches, so you can sleep soundly the whole night through.

The only remaining problem is that it isn't always easy to fall asleep in the first place, on account of the noise. By that I don't mean a lonely screech owl – they're usually quiet by midnight. No, I mean the beetles and other insects that rustle about in the branches. Unfortunately, they'll also be under your head, so you'll need to be truly exhausted before you finally close your eyes and fall asleep.

Survival in the forest is certainly not a walk in the park, and if you did need to feed yourself in the wild in an emergency, you'd find food very scarce. It is little wonder that in the times of hunters and gatherers, an area the size of Germany would have had a total population of barely more than 10,000 to 20,000. For large mammals (including us), forests are about as far as you can imagine from a well-stocked supermarket. You might have to walk long distances to find anything to eat. That's why wildcats have territories stretching over several square kilometres, for example, because in a smaller area they wouldn't find enough mice. The lynx, a larger cat, needs a hundred square kilometres to sustain itself. And what about us humans? While our ancestors would have needed at least ten square kilometres of forest per person, we're now squeezed into such close quarters that, depending on the country, our 'territories' are on average only about 0.004 square kilometres (4,000 square metres) in size. This is the average area of the territory in which we live and work, if we consider all the roads, train tracks, offices and shops, arable land and forests that we rely on to supply us and divide up the land area proportionally. This means we now house over 2,000 people on the same amount of land that once fed just one of our ancestors. For me, this is a fine example of how far we have moved away from nature. A little private survival adventure lets you experience just how far civilisation has already removed us from our roots and original tastes and preferences.

19. When the Forester Becomes an Undertaker

I WOULD NEVER HAVE DREAMED that someday we would be burying people here in the old beech forest I manage at Hümmel. And yet so far we have had over 4,000 funerals.

The trigger was the state forestry administration's plans to cut down some old trees and replace them with North American Douglas firs. So far, the community had successfully resisted, but I wasn't confident the question was ruled out for the future. After all, at the time I was still a state official, so if my supervisor ordered me to, I would have had to persuade the community to accept the felling of the old trees. And even if I had refused, what would it achieve long-term? What if my successor was less resolute than I had been and let the chainsaws run wild? Our ancient tree stocks would disappear like snow melting in the sun, because they are still being exploited to supply the timber industry with raw materials. Although Germany was originally covered almost completely by beech forests, today we have barely one-thousandth of the area of old, halfway-intact beech forests. And the Hümmel munici-pality, my employer, still has a hundred hectares of this ecological treasure, which is 15 per cent of the area of the entire forest (750 hectares). It was my deepest wish that it would stay that way.

Then, one evening, I was having a beer with a colleague after an excursion to the Black Forest and I learned from him about something curious going on in a forest in Hesse. They had recently started burying urns with people's ashes, my colleague told me. The forester there had been demoted to gravedigger! Hearty laughter accompanied this anecdote, but it made me sit up and listen. Could that perhaps be the salvation? If we transformed the forest into a burial ground and offered old beeches as living tombstones?

The very next day, I went to see the mayor of Hümmel and presented the new idea to him. Rudi was quite taken by the idea and so over the following months we set to work, carrying out our plan. We wanted to protect the old beech forest as it was with minimal disturbance to nature, so we wanted to avoid building any new roads or car parks. We repurposed an old wood storage area for parking and created the access from the old forest road, made wheelchair-friendly with a little gravel – and that was our burial forest finished. Well, almost – next, it was time to measure the girth of the trees. Each trunk had to be recorded to the nearest centimetre and logged on a map, allowing for up to ten potential urn graves around each one. The old beech and oak trees each got a small metal plate with a number, as well as a small plate the size of a credit card, which would list the names of all the people buried by that tree.

The first visitors were pleasantly moved and loved the idea of finding a place here for eternity. Less

sympathetic was the Catholic Church; many members
of the congregation struggled to come to terms with
the idea. The media soon became aware of this conflict
and so, involuntarily, we found our burial forest receiving
a considerable amount of free publicity in the press and
on the radio. What seemed to upset people was the idea
that the ashes would be absorbed by the trees' roots
and be brought back into the cycle of nature. Resurrection
was not possible, argued the representatives of the
diocese. But what about graveyard burials? The deceased
in their coffins also decompose at some point, which
means nothing other than being broken down by organisms, and thus returning to the cycle of nature. The
Catholic Church then amended its position so that
forest burials with a priest were officially approved by
the German Bishops' Conference. Meanwhile, the first
Catholic 'woodland churchyards' started to open, pipped
to the post by Protestant churches, which had started
to open such establishments years earlier.

I'm glad the turmoil of that time is finally over,
because it was mainly the relatives of the deceased who
suffered from the wrangling. If they desired clerical
assistance at a woodland burial, they had to ask either
an independent preacher, a Protestant pastor or a rebel
Roman Catholic clergyman. There was one such a rebel
in the early years of the Hümmel burial forest, who
could not bear to turn anyone down in their time of
need, or at least until he was suddenly transferred to a
new post one day. Today, nobody gets worked up about
burial forests; on the contrary, they have become established as a routine option in our burial culture. There

are now hundreds of such facilities across Germany, Austria and Switzerland.

So how does it all work? First of all, free group tours are offered in virtually all burial forests. If you like the forest and you're interested in reserving a burial space, you can arrange an individual tour. The forester will show you the remaining trees, respond to your personal requests and reserve your preferred spot. The lease contract is drawn up in the office and sent out for you to sign. A location plan completes the documents, and for a fee your space is reserved for ninety-nine years. At least, that is how long the contract is valid for, and barring any unexpected leaps forward in medicine, that time span should be sufficient for every adult alive today.

So, your loved one has reserved a spot, and eventually their time comes. Before the forest burial, they will need to be cremated, and for that you'll need to engage an undertaker, who will order a coffin (it could be anything from a traditional casket to a cardboard box, but yes – this is necessary for technical reasons) and make an appointment at the crematorium for a direct cremation; that is, one without a ceremony. Then you or another relative can pick up the ashes – at least, this is allowed in some federal states of Germany. Otherwise, the undertaker will deliver the ashes, or they can be posted to the burial site. Hmm, how safe is that, I hear you ask. Don't packages risk getting lost in the post? Well, I have never experienced it or heard of it happening. Maybe it's because of the prominent 'Cremated Ashes' label they stick on the front. So, now you know what to do if you have something valuable to post and you definitely don't want it to get lost ...

The urn has arrived, the grave is dug and decorated with green spruce branches. The funeral can now take place exactly as the relatives (or the deceased) envisaged. It can be very different to a burial in a traditional cemetery. For example, I recall a man burying his wife all alone in the snow-covered forest. Or the jolly send-off of one chap from Cologne who had asked his loved ones not to mourn him but to raise a toast to him at his graveside. So, his family and friends dragged a keg of local beer to the burial site and poured a glass for every guest. Of course, there are also ceremonies with a very traditional feel. Whether it's a funeral with a eulogy or a sermon, with music or poetry, it is important that everyone can have a farewell that meets their wishes.

Once the ashes are interred and the funeral guests are on their way home, the opening is closed over again. After just a few minutes you would no longer be able to distinguish the spot from the rest of the forest floor, and that's how it should be. After all, the burial forest is a protected area, and here nature is in the foreground. There is no visible grave to be tended, which saves the mourners time and money.

In the early days, I often heard people grumble: 'They bury you somewhere in the forest, then nobody comes to visit you again.' Quite the opposite, in fact: we get regular visitors. People come not with flowers and watering cans, but with the whole family and a picnic. Grandma and Grandpa get a visit for Sunday lunch just as they did in their lifetime, but now it's accompanied by a nice stroll in Hümmel Forest. The children can play, and the dog can come too, and afterwards perhaps coffee and cake at a pub or café nearby. There's nothing oppressive or gloomy about visiting this kind of cemetery.

And what about the ancient beech forest? Do the burials harm the trees? It's a tricky question and one I continue to ask myself. First, there are the graves: up to ten are dug around each trunk, eighty centimetres deep. This disturbs the sensitive forest floor, albeit only in certain small areas. We carefully lift out the soil by hand with a post-hole digger. Using machinery is out of the question, so as not to compact the soil and to avoid noise. After the burial, the soil is shovelled back in, as close as possible to how it lay previously. It cannot be exactly as it was originally, and besides, there is now also an urn in the grave. The plots are set at a distance of two metres from the trunk, so that no thick roots are damaged by the digging. If ten graves were to be laid directly at the base of the trunk, it would be tantamount to cutting it down.

On the topic of cutting trees down, every woodland burial supplier is liable for predictable risks associated with the trees. A rotten branch? The slightest breeze could bring it down and injure a visitor. A wounded trunk with a fungal infection? It could fall over and cause more than just a headache. It's all about public liability insurance and in some burial woods, the owners allow this to cast a shadow over their approach to woodland management, with overzealous foresters removing every potential danger to minimise the risk of injury and legal action. Every halfway unsafe tree is felled and the crown shredded into inconspicuous wood chips which are sprinkled on the many paths. But visitors want untouched nature, which is not disturbed by any sign of tree felling, or timber harvest as we should call it – because after all the trunks are sold to the nearest sawmill or as firewood. What was it we said about ashes returning to the cycle of nature? This cycle of tidying up high-risk trees might just lead indirectly to someone's ashes ending up in an item of wooden furniture, or back on the fire for a second time.

There are alternatives, but they're much more expensive. Here at Hümmel, all the trees are inspected several times a year, and if there is a potential hazard, we send the tree surgeons up. They climb up into the crown carefully with ropes and remove any dead branches. If an entire tree dies back, we don't fell it, we just cut back the crown. If left to nature, a dead trunk would crack down the middle, leaving a tall stump. The tree surgeons imitate this process by leaving the trunk standing, with the branches left to rot at its feet. This intervention is barely distinguishable from what would happen naturally; the only drawback is the considerable cost over the ninety-nine years. While in the low-cost version of a burial wood, the 'upkeep' can be financed by timber sales, for the eco-friendly, conservation-compliant alternative, we need to set aside considerable capital reserves. But which local government has a strong enough budget these days to set aside cash for a rainy day?

So, to answer the question, it just remains to say that, yes, the trees do suffer when, on the grounds of cost, the forest is cut back to resemble a park.

And the urns and their contents? Again, a thorny issue. After all, who wants to think about whether our loved ones' ashes might harm the forest? There's a fairly wide range of urns made from organic materials available on the market (yes, there is indeed a market for urns). Be it corn flour, wood or other organic substances, the result is usually a plastic-like product, which decomposes quickly in the soil, as the manufacturers claim. That's what I believed, at least, until four years after the opening of the Hümmel burial forest we received a request for an exhumation. An exhumation? How was that supposed to work with compostable urns? Anyway, the family wanted to move the remains to another location and the request was approved by the authorities, so my co-worker had to grab his shovel. To our surprise he unearthed one almost intact urn; only the coloured paint had started to blister. This was good for the family who wanted to relocate the urn; bad for the forest, where we wanted to leave a minimal impact on the soil. Since then, we have stipulated that only containers made of untreated beech wood fixed with natural glue may be buried under the old trees. The glue dissolves with moisture, so that the urn can break down and the ashes can return to the cycle of nature. The wood itself may need some years or even decades to decay, but it behaves like dead roots, so is completely harmless to the forest.

And the ashes? Many people believe that the contents of the urn are mainly the ash from the coffin anyway. In fact, it's the other way around. The bulk of the ashes comes from the body (mainly the bones) and only a tiny proportion is from the wood of the coffin. So far, so good. But if the deceased person was sick, what about contaminants? After all, we absorb so many chemicals when we take medication; might some of the contamination of our tissue be carried over into the ashes in similar concentrations? According to the crematoriums, this is not the case. Toxic heavy metals such as mercury are filtered out with the extraction fan (goodbye amalgam fillings!), leaving behind almost pure chalk from the bones.

There are those who would refute this statement, however. Several years ago it was argued that the carcinogenic compound chromium-6 was formed during cremation and could lead to ground water contamination. Woodland burial operators investigated the claim and found it was linked to certain individuals with a conflict of interest. After all, woodland burials risk making entire industries redundant. Cemetery gardeners and stonemasons risk being pushed out of the market, not to mention coffin manufacturers. Wouldn't it be convenient if these natural burials went out of fashion? And yet there remains a whiff of doubt. Even burning pure wood forms this toxic heavy metal compound; however, it degrades relatively quickly to a harmless variant in the open air, according to current understanding. I will follow further research with interest, because, after all, beside the risk of water contamination, my concern is for the protection of the old trees, which mustn't be pushed into the background in the debate about urn burials.

The decoration of graves or tending to them in any other way is not permitted in the burial wood at Hümmel. The only thing that can be placed into the grave during the burial are natural things like pebbles or shells, mementos from a shared vacation, perhaps. During the ceremony, the surrounding area can be decorated with flowers and candles, and music can be played – as with any conventional funeral. When the relatives have left, my co-worker collects the flowers and brings them to the small place for offerings at the entrance to the wood.

Incidentally, the history of burials has more to do with the forest than one might think. Coffin burials came into our culture together with Christianity. In earlier times, the Germanic tribes and the Romans burned their dead, the latter using urns for the ashes. However, Christianity has its roots in the Middle East, where wood is in short supply. You needed to burn several cubic metres of wood for each cremation, an unaffordable expense when there are few trees available. A coffin which used only a fraction of the precious commodity was far cheaper. With the new religion, we inherited the custom of burials, which in fact made less sense here. In this respect, burying ashes after crema-tion is little more than a return to a formerly widespread tradition of forest burials – a thought I find appealing.

What do people experience when they come to the burial wood? This question has become the most import-ant for me. Above all, the woodland setting conveys one thing: peace. I've often heard visitors say that they walked down the path from the car park through a spruce forest to the old beech trees and when they were walking beneath the old deciduous trees they suddenly felt at home. I don't know quite why, and even they couldn't fully explain it. Perhaps it's the sense of being in an intact forest, which enjoys a certain equilibrium within itself and with the environment. Maybe we humans still have a buried, yet not quite forgotten, instinct for whether an ecosystem is healthy or disturbed (as with the spruce plantations). In the distant past, such things may have been important, because an intact forest ecosystem offers more security against storms and contains greater food reserves.

Choosing the right tree is often a cheerful affair, at least when it comes to reserving a spot just in case. You hear couples joking about having a test lie-down or laughing as they imagine the eternal rounds of cards they could play here together one day. Women often fancy a spot in the sun because they've always got chilly feet; men are sometimes drawn to the trees near the brook because they're keen anglers.

The act of choosing a tree can feel liberating for many people. I once brought an old couple, both well over eighty, into the burial wood. Both were so ill that they had only a few weeks left to live. Walking was out of the question, so I drove them down along the main track in my 4x4. We drove at walking pace past the huge trunks, and when we stopped at the mightiest mother beech in the entire wood, they both immediately felt in their hearts that this was the place. They bought two burial plots, and when we got back to the car park, they told me, 'This was the best day out we've had for a long time!'

Even after the funeral, it's not unusual to come across novel ways of dealing with grief. I once came across a woman sitting under the tree where her husband had been buried. Here she sat, with a happy smile, writing poems in the bright spring sunshine. On the upper slope of the forest, a motorcyclist can be seen once or twice a year. He sits quietly at the grave of his friend who died in an accident, downs a bottle of beer in his honour and disappears shortly after.

And then there was the ice. It certainly isn't rare to see chunks of ice on the forest, but on a hot summer's day in July, I had no explanation for my find. I puzzled for weeks over what it could have been. An especially cold summer's night with frost? It can happen in exceptional years here in the Eifel region. Or could someone have lost something from their freezer? This turned out to be closer to the truth, but the true story is much more touching. It was a lonely man whose wife was resting in the Hümmel beech wood. Grave decorations are not permitted, or desired, and the widower respected this. But, the creative soul that he was, he found a way around it. He made some heart-shaped moulds at home, poured them full of water and froze them in the freezer. He took the ice hearts with him to his wife's grave and let them melt there in the summer sun.

For me these experiences are consoling. Because in the beginning I wasn't sure if I would be able to endure long-term all the grief that comes with funerals. After all, it's not so common that a thirty-eight-year-old forester (as I was then) deals with death on a daily basis. Today I know that the atmosphere at the burial wood helps people to cope with their grief, and that gives me a good feeling, the sense that I am doing something positive.

20. Is That Allowed?

Who owns the forest? When I ask primary school pupils, the response I often get is, 'You!' That's not true, of course, but it seems most people have lost the sense that at least the public forests in this country are communal property. They actually belong to every one of us, at least in part.

Fifty-six per cent of the forested area of Germany is in the hands of municipalities, cities or the state. This accounts for around 800 square metres for each resident of Germany, and the proportion is even higher in Austria and Switzerland, with 3,800 square metres and 1,150 square metres respectively. On each of those 800 square metres, according to the German Federal Ministry of Food and Agriculture, there are on average over a thousand trees, or saplings. More than 99 per cent are very young; genuinely big trees need 400 square metres each. Whichever way you look at it, there's a considerable patch of forest for every citizen of this country. So that the forest is managed or protected as a large ecosystem in an orderly manner, and everyone doesn't do what they like with 'their share', the principle of democracy applies. The course set by parliament is implemented by forestry administrators at a national and regional level – your service providers paid for by taxpayers' money.

Why am I stating the obvious? It's because of a growing sense I have that these service providers are not always aware of their role. For example, there is a consensus among German political parties that 5 per cent of forest land should be protected by 2020. The trees should no longer be used for timber on these areas; this should be land where nature is allowed to take its course. Since almost half of the forested land is privately owned, it is generally agreed that these protected areas would be primarily made up of publicly owned forests. That would mean that 10 per cent of the public forests would be taken out of use for logging. Now, you would think that this would be a doddle to implement, but you'd be surprised. So far, not even 2 or 4 per cent has protected status, and that won't change quickly. The reasons are among others the state forestry commissions, who sing from the hymn sheet of protection by commercial use. Every square kilometre that is no longer available for harvesting would increase the pressure on tropical forests, so the argument goes, since the wood would then have to be imported from there, resulting in just as much forest destruction.

Apart from the fact that we could simply consume less wood, I think the argument is a touch colonial. Besides, it's high time citizens got more involved in caring for their forests. You can influence what your state-employed forester does in the woods on your doorstep, after all. One example of this is a ground-roots project called the 'Friends of Königsdorf Forest' (waldfreunde-koenigsdorf.de). This group campaigned to rescue a nature reserve, which had been established to preserve an old deciduous forest. Despite the protected status, old trees were still being felled there and the heavy machinery was damaging the soil, so that in effect the forest hardly differed from other commercial forests. The citizens did not want to sit by idly and watch this happen, so they began to get involved and get vocal. A few years later and after lengthy dialogue with politicians and the press, the citizens' initiative has become a serious stakeholder that has had a strong influence on the fate of the nature reserve. Other examples can be seen in other regions, showing that even small groups can make a difference as soon as they engage with the public.

What about the other way around? What are you allowed to do beyond what we've already discussed? As we've already covered, you're allowed to hike anywhere in Germany, including off the beaten track, and you may pick wild mushroom and berries, but what about camping in the woods? In Scandinavia, there's the universal right to roam, as mentioned earlier, which also allows everyone the right to camp one night in a tent even on another person's land, the only exception being the immediate plot surrounding a private house. This is different in Germany and necessarily so, because with our much higher population density we would have no quiet space for wild animals to retreat to if we could camp anywhere. That's not the whole story, though – at least not for the approximately two million private forest owners and their families. Any time you spend on your woodland plot is considered forestry work. And if you work outside in all weathers, you have the right to make a fire to keep yourself warm. Never mind the forest fire threat level, you can always light a fire for a barbecue in your own forest. Depending on the state, it's also permissible to stay overnight in the woods, if the property doesn't fall within territory classified as a special protected area (such as a bird sanctuary). You can also extend this permission to others.

However, none of this is permitted for private companies; commercial operations have no right to pick mushrooms or berries, to camp or even to use regular woodland tracks. And I think that's appropriate. In a country as densely populated as Germany, landowners have to tolerate a lot of coming and going on their land, but their social responsibility to allow everyone access to nature is part of the justice of a modern state. But if you wish to profit from someone else's property, you must ask for permission and in certain cases pay for the right. This applies not only to outdoor events such as survival training and demonstrating new off-road vehicles; it also covers horse-drawn carriage rides or fun runs.

It makes no difference whether it's individual leisure activities or organised events, with ever greater frequency we hear about the question of liability in the event of an accident. Forests can be dangerous, after all, and it's not the wild animals that make headlines, but the trees. Don't worry, there are no dangerous trees per se, but if an elderly giant loses a dead branch when you're walking beneath it, you'd know about it. From a height of forty

metres and with the weight of a water cooler, the force
of a blow from a falling branch would be such that an
accident of this kind could be fatal. A broken piece of
wood on a bike path might also make for an unpleasant
surprise. No wonder that legal action is becoming more
common, and these events can be particularly unpleasant
for the forest owners and foresters. For if a walker suffers
physical injury or even death, then public liability insurance doesn't even apply as this wanders into the jurisdic-
tion of the criminal courts.

The general legal position is that the forest owner
is liable for hazards arising from his property that
pose a threat to the general public. The law doesn't
specify much more than this, so in this regard, all woodland owners have to go by is precedent – that is, previous
court judgments. The problem is that attitudes are
changing, so there is no definite certainty as to how to
proceed correctly. Is it enough to inspect all the trees
that line a road twice a year? And if so, what information should be logged? What qualifications are needed
by those carrying out the inspections? Safety is certainly
taken very seriously. But is it the safety of the general
population that is the priority? Sometimes I have the
impression that it is more about the safety of those who
would be held accountable. Because no one wants to
take a risk, and where there is any doubt they clutch at
large-scale felling as the solution. You'll often see a road
through a forest where every single tree lining both
sides of the road has been cut down. This rules out the
possibility of a rotten tree being overlooked and falling

onto a passing car. According to the Federal Statistical Office, there is a total of 230,000 kilometres of roads in Germany's regional road network.[26] Besides this, there is another 600,000 kilometres of small local roads, over 33,000 kilometres of unmade tracks, and tens of thousands of settlements edged by woods. If these radical precautionary measures were carried out everywhere, we would hardly have any woodland left.

How great is the potential danger? I have not found any statistics that show the annual figures, but the industry magazines report every case involving a lawsuit to determine whether or not a landowner can be held accountable. There are only a few isolated cases in a decade where the blame can be traced back to a problem tree that should have been identified as rotten. The far greater number of life-and-limb injuries are caused by storms, where sometimes entire woods are blown over, burying roads and cars. It seems disproportionate to me, therefore, to clear thousands of kilometres of strips of forest and to fell every single tree that's anywhere near a country lane and has the slightest sign of a woodpecker nest hole. But sometimes, with such harsh measures, two birds are killed with one stone: there's no risk of a tree falling onto the road, and at the same time huge amounts of wood are produced, which can − very conveniently − be transported to the nearest biomass power plant.

21. The Forest at Night

WHERE DO YOU FEEL MORE COMFORTABLE: in the pedestrian zone of a big city during the day or at night alone in the dark forest? And because you already know what I'm getting at, I'll just come out and say it: you don't know until you try it. Our senses and our instincts are on high alert, when all we have to take in is a few strange background noises, and nothing familiar. Might there be something lurking in the gloom? Was that the sound of twigs cracking, as if a big animal were approaching? Even I still occasionally experience a slight sense of unease, although I know that nothing can happen to me. Our senses deceive us, amplifying the feeling of threat, perhaps because of the genetic heritage we carry from our ancestors. Whereas once gangs of bandits lurked in the woods, or even further back in time, sabre-toothed tigers prowled amidst the trees waiting for easy prey, today, statistically speaking, the forest is one of the safest places we could be at any time of the day. Which pickpocket lurks behind a tree to ambush a passer-by? In most woodland areas, he would waste away before a worthwhile victim showed up. The city centre would be a better bet.

No, a forest at night is a particularly safe and beautiful experience. As the light fades, so too do the sounds of civilisation. The rush-hour traffic has died down, the lawnmowers are silent, the construction site has stopped for the night. The only sound to greet you is the occasional plane overhead. Why does this matter in the forest? Because when it is really quiet, you start to realise how far sounds can carry. And for an unadulterated nature experience, you need natural sounds. I know now from working with camera crews how hard it is to block out the sounds of the world beyond the forest. They like to record a track of 'ambient sounds', i.e. a few minutes of boughs swishing and birdsong, to play in the background of scenes without dialogue. These sequences are so important, because usually everywhere we find ourselves, we barely go a single minute without acoustic interference of some kind, usually road or air traffic.

So, if you want an uninterrupted natural soundtrack to your nocturnal hike, there are two options. One is to descend into a mountain valley. Because mountains block out all noise from beyond, uninhabited valleys are wonderfully quiet. You may still get the occasional night traffic overhead, though. The second option (and much easier to accomplish) is a forest walk on a breezy night. When the wind whistles through the leaves and twigs, when branches rustle and arching trunks groan, this soundscape overlays all other acoustic sources. Now you have the most perfect symphony that a nocturnal forest can provide. Windy conditions allow you to hear exactly what thousands of generations have heard before us, the sounds that would have made up the background soundscapes to countless Stone Age campfires. Walking in the night like this gives me a certain sense of freedom and of timelessness.

For unrestricted enjoyment, it's best to stay on the footpath. Otherwise, your walk quickly becomes about as much fun as a poke in the eye – quite literally. Particularly in a coniferous forest, there are many thin, broken-off branches that protrude low down on the trunks, which pose a high risk of injury. Still, I would resist the urge to pull out a torch. Artificial light immediately catapults you back into civilisation and brings your primeval fears to the surface. Everything outside the beam of light becomes so much harder to see. Not only that, your eyes will have been adjusting to the dark and switching on a torch knocks this process back several hours. Your eyes can adapt perfectly well to night vision, just very slowly.

The retina has two types of photoreceptors: rods and cones. During the day, it is the tiny cones that process the light entering our eyes. These only function in well-lit surroundings, as is the case in the daytime where there is plenty of light outside and in brightly lit rooms. At night, however, the cones can't cope with the low level of light and instead the rods come into play. It takes a while for them to take over, and when they do their functionality is more limited than the rods. They can't process colour, so the world appears in black and white, which, by the way, explains the old saying 'All cats are grey in the dark.' And because it is much darker in the forest than out in the open, it is worth giving your eyes time to adapt. If you need to switch on an artificial light, it is best to use red. Red light doesn't shift the eye out of its dark mode, and this is the reason why astronomers, for example, use red bulbs when they're observing the night sky with a telescope. To find a hand-held red torch, you could look online for a red stargazing flashlight or use the cycle lamp from the back of your bike.

Another way to see something in more detail is to choose the right occasion. For your first night hike experience, a full moon on a cloudless night gives the ideal conditions to help you find your feet. The full moon is even bright enough to read the newspaper outside.

We have focussed for a while on our sight, which doesn't supply us with very much information in the dark. We pick up a lot more with our ears. Large mammals give themselves away with rustling sounds underfoot, but only when it is dry. As soon as rain drenches the floor covering of twigs and leaves, it becomes soft, soggy and almost silent. Luckily, the animals themselves also emit sounds. If you hear a bark, it's not dogs in the deep forest at night, but deer. Their hoarse call is known in German hunting jargon as *Schrecken* ('alarm' or 'terror'), because the animals are precisely that: terrified. On the one hand, they bark to warn their peers, but on the other hand, they risk attracting potential predators to themselves. Therefore, they will only sound the alarm if the disturbance is several hundred metres away. This leaves enough time for all the deer to retreat inconspicuously. If a deer is surprised by something at close range, it will jump up suddenly and make a quick, silent getaway. It often happens if you're cycling or riding on a quiet lane: some deer might be dozing on the verge and won't notice you coming until you're pretty close. At such close range, the surprise is the same for both parties and gets your pulse racing – and theirs.

And what about our noses? Well, this is not our strongest sense, but if you can see very little, then every impression counts. Besides, we perceive scent more strongly when our concentration isn't distracted by images. Perhaps you're just getting used to the forest ambience. The soil is so densely colonised by fungus that it pervades every millimetre. Especially when it's wet, it makes for a marvellous smell. Or how about the aromatic scent of the conifers? Like a mixture of resin, orange peel and sugar, it brings back memories of summer holidays in the Mediterranean, where the air is full of that piney scent. It's a composite aroma made up of various components and it sends a scent signal. Many conifers suffer because they are too hot and too dry. They easily fall victim to invasions of bark beetles, as they can barely defend themselves in their weakened state. To warn their comrades, they exude olfactory cries for help – and to us they give off that holiday aroma. The trees also clean the woodland air: the scent they emit also contains germicidal substances that help keep fungi and bacteria at bay.

On your night hike, don't be deterred by 'No Entry' signs hung out of alleged concern for the game animals.

These are usually put up by hunters, who just want some peace and quiet but use the needs of the deer as a pretext; it's more persuasive, after all. Whatever they say, in Germany your right to roam applies at any time of the day or night, even in private forests.

So, what actually happens in the forest at night? Is that when everything comes to life? Not quite, at least not as far as the trees are concerned. They fall into a deep, restful sleep just as we do overnight, to take a break from day-to-day business. The process of photo-synthesis comes to a rest and all the movement within the trunk and crown winds down for the night. This also impacts on the oxygen content of the air, because trees also burn sugar and other carbohydrates, using oxygen in the process. They are by no means just oxygen producers, a function to which they are often reduced, along with providing wood. No, they also breathe in oxygen through hundreds of thousands of tiny mouths, the so-called 'stomata' on the underside of the leaves and needles. During the day, trees produce a surplus of oxygen, which results from their breaking down water and carbon dioxide and converting these to sugar with the help of sunlight. At night, however, the giants draw on their carbohydrate reserves under the bark and release CO_2 in the process, just as we do. The healthy forest air is somewhat less healthy at night, but by so little that in practice it hardly makes a difference.

There is one new insight science has given us which I find particularly touching: when it gets dark, the trees literally curl up to sleep. A research team from Austria and Finland scanned the crowns of birch trees and were surprised to find that as soon as it got dark, the trees let their leaves and twigs droop, and they drop even further throughout the night. The difference in height between the night and a bright sunny day could be up to ten centimetres. It is still not clear whether trees are woken up from their slumber in the morning by the rising sun or by an internal clock.[27] There's another process that you wouldn't notice either: at night the tree trunks grow fatter. Just a tiny bit, but enough that researchers can measure it. The reason for this is that in the night the water that flows up into the trunk from the roots is no longer drawn into the leaves, as they're in a dormant state.[28] When production picks up again with the first rays of the morning sun, the water belly disappears.

In the case of animals, their activity increases as dark falls, because those which are hunted, such as deer, no longer sense danger from their human predators. Other species such as bats are nocturnal specialists anyway and hunt moths by ultrasound detection. It's thought that the reason these moths are furry, incidentally, is to try to outwit their predators. The rough surface on the body and wings muffles the sonar signal and makes it harder for the bats to detect. Apart from that, the moths have developed an excellent ear for the highest pitches and can hear the bats out searching for them.[29]

The nocturnal flight of the owl is endlessly fascinating. Owls look out for mice, but also for other birds roosting in the trees. Their feathers have soft, fringed edges, making their wings beat in absolute silence. Like nocturnal phantoms, they appear and then they're gone again. Their prey remains unaware, sensing their fate only when it is already too late.

If you think you're not being watched by other people at night, then unfortunately I have to disappoint you. In Germany, whenever you pop behind a bush to relieve yourself, there's a good chance you're being observed by a video device triggered by an automatic sensor. Everywhere you go these days in our woodlands, there are small, inconspicuous video and still cameras attached to trees, which capture images and videos as soon as the motion sensor registers your presence. Now, I don't mean to suggest that anyone is intentionally collecting voyeuristic pictures of walkers attending to the call of nature. No, the intended subjects are the local deer and boar populations. The images are recorded with a time stamp so that game hunters work out the best time of night to climb up on their high seats and sit watch with their rifles. This way they save themselves long nights shivering out in the woods waiting for prey to show up. Once they establish the time their targets tend to come by, they can be ready to shoot the second they show up.

These wild game cameras seem to be multiplying like rabbits. They are cheap to come by online and they're often discounted at the large, low-cost grocery stores. Every little helps... At less than a hundred euros each, these allow well-heeled game hunters to control large parts of the wooded wilderness by installing them in every corridor of the woods that game animals have to pass along. Termed *Zwangswechsel* in German hunting jargon, these paths are like narrow bottlenecks, routes that game animals are forced to follow by obstacles such as dense undergrowth, a steep slope or a marsh. If you have cash left for more cameras, other favourite spots for installing these devices are salt licks, game pastures or places where fodder is left intentionally. Bottleneck paths? That means you too have to pass along these corridors as you move through the terrain (you'll have the same problems making progress through the thicket or other obstacles as the wildlife). And pastures are also nice places to stop for a rest on your walk. Will you feel so relaxed next time you stop for a picnic, knowing there's likely to be a camera, hidden away at the edge of the meadow amongst dense branches, filming and photographing you mercilessly?

Does it sound like I'm exaggerating, maybe even paranoid? The data protection officer for Rhineland-Palatinate, Edgar Wagner, thought otherwise. He expressed his views on the matter so clearly in 2014 that even the major broadsheets paid attention and finally realised that some 100,000 game hunters had installed a similar number of wild game cameras on trees trunks across the country.[30] He argued that the forest is a public space and its surveillance by private individuals is illegal. Violations of public privacy in this way can be punished, at least in Rhineland-Palatinate, with a fine of €5,000. And has this had a knock-on effect? Sadly not, and I must say that since the advent of wild game cameras, I've felt like I'm on *Big Brother* – constantly being watched everywhere I go. My work means I'm always out and about in the forest on foot, and I don't like the idea that reams of images are being taken of every move I make.

There's another group of people for whom this surveillance is even more unpleasant: adulterers popping into the woods for a little hanky-panky, perhaps. In my thirty-year career, I have only twice stumbled upon couples in the act, and that was only by chance. They leapt like they'd been stung by a tarantula as soon as they heard my approaching footsteps and hid behind their car. I had simply wanted to see who had parked their car on the verge and felt bad for embarrassing them. The situation was quite different for an Austrian mayor who was filmed during an intimate moment by a wildlife CCTV camera and the pictures were published. He is not alone in his predicament; several of his colleagues in Austria and Germany have had the same treatment.

Whatever happens with the footage, I would be glad if it were finally made illegal nationwide to install this sort of monitoring equipment in a public space. Our woodlands are one of the last retreats in our populous landscape and they should be left that way.

22. Dress Code

THE OUTDOOR CLOTHING INDUSTRY IS BOOMING. If you look at the catalogues, there's so much choice, it is difficult to decide which trousers, shoes or jacket to go for. Even for someone as tall as me, there's too much choice and I need reviews to help me navigate the options. And browsing in a shop doesn't always help, although at least you can try things on. But we're less likely to get it completely wrong if we follow a few basic rules. The simplest is: look at what the pros wear. People who are out and about all day can't compromise on quality when it comes to clothing.

Have you ever wondered why foresters always seem to wear green? A hundred years ago, the answer might have been because of the poachers. Here and there you'll still find a lonely memorial stone in the forest, bearing witness to my predecessors' heroic struggle against poaching. If a forest warden was killed in service, a memorial would be erected at the place of his death. Good camouflage could be crucial. Today, it's the other way around: if you're too well camouflaged, you risk ending up under a falling tree, if you're involved in felling, at any rate. Lumberjacks work in groups, and if you can't see your colleagues, you might set the chainsaw in the wrong place. All workers must wear high-visibility orange patches on their safety gear at the very least. Foresters, however, tend to work alone as they make their way among the trees, deciding which trees should be felled and marking the trunks accordingly with spray paint or tape. This inevitably leads to body contact, especially with the broader trunks, which looks to a casual observer like a tree hug. This close contact leaves algae-green traces on your clothing, but if you wear olive-green clothing it's less likely to show up.

Incidentally, it doesn't make any difference to the deer and wild boar what colour you wear. If you think just wearing green will make you so hidden from view that they won't see you, I'm afraid you're mistaken. Much more important than the choice of colour is a pattern that breaks up your outline. The best camouflage means you are no longer visible as a large 'animal' but blend in with the different blocks of light and shade in the undergrowth. This is exemplified by the tiger with its vertical stripes.

Deer hunters find themselves in a quandary: on the one hand, they need to be visible in accordance with the law, dressed in high-vis orange, for example. On the other hand, they want to be camouflaged so they can see and shoot as much game as possible. The clothing industry's cunning response was a camouflage-patterned, high-vis jacket. Does that sound crazy? And yet it works. My former boss, a forestry commission officer, was once out deerstalking when he was almost run over by a roebuck, which realised only at the last second that he wasn't a bush. The reason for this confusion? Large

forest mammals are partly colour-blind and can't distinguish reds from green or yellow. So for the roebuck who
ran into my colleague, the camouflage pattern made the
jacket blend in with the surroundings. The only exception is the colour blue. Strictly speaking, red deer, roe
deer and wild boars can only distinguish between blue
or not-blue, like most mammals, so you have a wide
choice of colours for your future wildlife-spotting
jacket.[31]

After colour, the next question to consider is fabric.
Personally, I'm not a huge fan of high-tech artificial
textiles when it comes to jackets. Yes, they're good at
keeping the moisture out when they're new, but they
often crumble and fall apart after only a few years.
I prefer clothes that serve you reliably for half a lifetime.
I will concede to the benefit of a mixture of cotton and
synthetic fibres, because this combination dries quickly
and is nevertheless very robust. Jackets and trousers
made of this kind of sturdy fabric won't get damaged
by a walk through the thorny thicket, and if the jacket
is thick enough, it should be fine even without a waterproof coating. It takes at least an hour for the water to
penetrate, and in most cases you will have found shelter
before then, for instance under a mighty old spruce.

In really bad weather, sometimes the best hiking
boots don't help, because water finds its way through
breaks in the membrane, in spite of all the promises.
What happens with jackets applies here, too: where
there's a kink, the fabric can break easily and will eventually let in as much moisture as leather boots made

without high-tech textiles. Either you need to replace them on a regular basis, or when it's really wet you just switch to good old welly boots. But which kind? Not the cheap version, which are made of plastic and are hopeless in winter. They are rock hard in the cold and you'll be slipping about on the frozen ground. Cheap wellies are also not exactly the healthy choice for your feet, because they rarely fit properly. You can walk better in genuine rubber boots, which remain supple even in the hardest frost and often have a better moulded insole.

We have already touched on the subject of trousers when we talked about ticks. The fabric should be light and unpatterned so you can easily spot the little black critters. Light beige, light olive green or anything like that works well, while also being a colour that doesn't show the mud that inevitably splashes up from your shoes to your trouser legs – after all, you'll want to be able to stop off at a country pub on the way home from your walk without looking like a woodland troll. On that subject, picture the scene. Your stomach's starting to rumble and happily you stumble across a cosy-looking inn. Now you could just stroll in and order your food, but after your rainy, muddy walk, you'll leave a trail across the floor that would make Hansel and Gretel green with envy. After an hour at the table, all the mud stuck in your boot treads will have come loose and will be clearly visible on the floor or, if you're in worse luck, on the carpet. It's happened many a time to me and it's always rather embarrassing, even if the landlord tolerates it with good humour (or doesn't even notice). Better to deal with it before you go in and to clean your shoes in the forest. Luckily, nature kindly provides us with plenty of cleaning materials. In rainy weather (and it's only then that this muddy problem arises) the streams and ditches will be full of water. Here you can swoosh your boots back and forth a few times in a shallow stretch to rinse the mud off. As long as it's only for a minute or two, even leather boots without any water-proofing will withstand a short wash.

If there's no running water, you can wipe your boots on a wet tuft of grass. If you shuffle back and forth a few times (don't forget the back of your heels!), your boots might not be as good as new, but at least you won't have kilos of mud hanging off them. And if there's not even a tuft of grass where you are? Then an energetic shuffle through the undergrowth will help, where the plants and twigs can be your shoe brush. Then to get them truly spick and span, you could wipe off the last smear with a bit of moss. On a rainy day, this soft green pad, drenched in clean rainwater, makes for a good wet wipe for washing your hands.

23. The Forest at Home

I CAN SAY WITH ALMOST COMPLETE CERTAINTY that a forest once stood in the place where you're sitting reading this book. How can I be so sure? Before humans began to turn the entire landscape upside down, there were virtually no treeless areas. The only exceptions were riverbanks, where floods and drift ice would repeatedly tear out old trunks, or large swamps and bogs. And, of course, the few areas above the tree line, for example in the Alps. But that high up, you're unlikely to be reading, so my guess is you're sitting where a forest once stood.

Our ancestors saw the dense forest as a threat because it provided little food and was where enemies lurked. Predators or bandits – either way, you didn't see them until they were mere metres away. So, what could be more tempting than to remove the troublesome hiding places and reap both tonnes of wood and acres of farmland at the same time? By 1800 this project was complete: large parts of Central Europe were like a steppe, the grassy ecosystem from which we originate evolutionarily. Great! But this joy was always mingled with a touch of nostalgia. In losing its trees, the landscape had lost its soul. The melancholy landscapes of the painter Caspar David Friedrich date from this period, with their gnarly oaks stretching their bare branches into the moody sky.

And where did our forests end up? The trees went straight to the nearest sawmill, just as they do today. Over 98 per cent of the land is in regular commercial use; that is, where the trees are not granted the opportunity to grow old. Apart from the fact that the proportion of land given over to conservation could be larger, there is nothing inherently wrong with the commercial use of timber. This natural commodity brings a little woodland atmosphere to our interiors, and current attitudes towards wooden furniture reflect that. What used to be considered an unacceptable fault in wood is nowadays emphasised as a feature. Imperfections such as knots and swirls in the timber, colour deviations and even wormholes – these all make a piece of furniture unique. There's a certain way of cutting timber so that you can feel the annual growth rings, making a desk, for example, something you can experience with all your senses. And the closer you look at a piece of furniture, the more you can discover about the life of the tree.

You might see, for example, thin, short lines – records of hard drought, which tend to be very faint in wood from deciduous trees and more pronounced in evergreen timber. There could have been splits and cracks in the trunk, which wouldn't have been without pain for the tree. These are usually caused by strong winter storms, which can bend the wood with the equivalent of a tensile force of up to a hundred tons.

If the log is sawn longitudinally into boards, the growth rings show on the wood as the grain running along the length of the board; that is, if the tree grew straight. But if something prevented the tree from growing straight up, swirling patterns will show in the boards, resulting from the tree's efforts to repair and steady itself. It could be that a spruce grows crooked at a young age and then compensates for this later on by growing more wood on the other side. In the sawn board, this will result in the streaks running at an angle across the board. Sometimes it is an injury that results in twisted or crooked growth. A tree is knocked over in a storm, for example, and scrapes off some of the bark of its neighbour, causing a severe wound. In order to prevent the invasion of wood-destroying fungi, the battered tree tries to close the wound up again through particularly rapid growth, resulting in a wooden lump, which can be very large depending on the injury. Bad for the tree, good for the carpenter: it makes for an interesting, varied pattern on a table top.

An encased knot also reveals something about the life of a tree. If it is the same colour as the surrounding wood, then this stump of a broken-off branch was still green (i.e. alive) at the time the tree was felled. The knot was fully healed over and does not pose a problem visually (well, it's a matter of taste), nor in terms of the wood's durability. It's a different matter when a knot is edged with black or where the entire knot is clearly darker than the rest of the wood. This would have been a dead branch stump, which the tree was in the process of filling in when it was felled. The branch stump was not completely surrounded by new, healthy wood; the tree was felled before it could complete its repair work.

If the knot on a plank is circular, then the saw cut across it at 90 degrees to the direction of growth of the former branch. And since the branch stump was already dead, it isn't firmly connected to the surrounding tissue. When the plank dries, the knot shrinks and falls out as a disc. The result is the familiar knothole that you can see through. It makes a charming feature, but is generally less desirable in furniture or floor boards. In the case of high-quality manufacturers, these flaws are identified during the production process and the knotholes are filled with a wood plug of the same tree species so that it is almost invisible.

If boards or furniture don't have any knots, it is likely that the wood is from a particularly large, old tree, which lost its old, dead branches long ago (it discards its low-lying branches that it no longer needs below the crown). Its branch stumps were covered over with thick layers of new wood decades ago. 'Knotless' timber tends to command the highest prices.

With certain woods, you can even look at furniture and determine the age of the tree when it was felled. Take beechwood, for example. There was a time when the preference was for light-coloured, flawless wood. This meant that older stocks aged over 140 years diminished in value, because the core of the trees develops a reddish discolouration. This red 'heartwood' means that when you saw up the tree, no two boards look alike; the wood shows colour deviations and even flame patterns. Fortunately, the furniture industry has responded to lobbying from foresters and has for years now been selling beech red heartwood under various pseudonyms such as wild beech or core beech. Nowadays people are happy to buy this more mature wood. The good news for the trees is that they can stand a few decades longer in the forest and age with dignity. In Germany, black storks can build nests in their mighty crowns, and woodpeckers can make use of a cavity here and there. And because as they get older, individual beech trees die off, with time there's an increased proportion of deadwood within the older stands, which is important for wildlife. So, if you want to help the birds, insects and fungus, red core beech is the wood for you. Or look for furniture made of old, mighty oak trees, centuries-old fir or larch. And to ensure that the forest, from which the wood originates, is managed at least halfway sustainably, look out for the FSC seal – just as you would with the firewood.

Your influence is even more direct when you buy a table and chairs from a local carpenter. There are real pearls of craftsmanship among them. I had my new desk built by a small local business, though I have to admit, it came about by coincidence. With my height of 1.98 metres, I was unable to find a suitable desk at one of the big furniture chains. For the good of my back, I knew I needed to have one custom-made. Then a small company signed some employees up for one of my workshops, and the name of the company made my ears prick up. Holzgespür ('a feeling for wood') is a bespoke joinery where customers are involved in the production right from the start. Ordering a desk from them, I got to choose from a range of locally sourced timber. Would I go for a wood with a lively growth-ring pattern or one with a straighter grain? Would I like to have knots adorning the top surface? To make it easier to choose, the company director even sent me a short video, which took me on a virtual tour through the workshop so I could inspect the timber in detail. They also kept me updated several times over the course of construction, and when my desk was finally installed in my office, I was delighted, because it far exceeded my expectations.

Bespoke furniture is of course more expensive than a mass-produced budget item, but the solid construction and the timeless natural design mean you get a real future heirloom. A shift from our disposable mentality in our home and office benefits the forest again, as it means a reduction in wood consumption. And we consume a lot! In Germany alone, we get through more than 150 million cubic metres per year,[32] which means a similar number of felled trees. But our forests fall a long way short of this demand because, according to the Federal Statistical Office, less than sixty million cubic metres are felled annually in Germany. While the forestry industry claims it would be no problem to raise logging quantities, conservationists already consider the current level problematic.

There is one very special tree which we bring home whole, and place in our living rooms at Christmas. The German custom goes back to pre-Christian times. Evergreen plants such as spruce and pine, but also yew and holly, were symbols of the return of spring. In the modern sense, the first Christmas tree with edible treats hung on its branches probably appeared around 1419, when a baker in Freiburg decorated a spruce tree with sweets.[33] The custom became properly established in the sixteenth century, but only among the well-to-do. It took another 300 years for conifers decked with candles to find their way into every household.

And what do the trees themselves have to say about it? We don't know, and it can't be much, because if they are mounted in a stand, then they're already dead. A more compassionate approach is to buy a spruce or fir with its root ball still attached. Then, when the holidays are over, the tree can be planted in the garden to carry on living there. I find this a touching gesture, even if the consequence is a conifer that rapidly outgrows its welcome. Keep your eyes peeled for huge blue spruces in the front gardens of houses. Why blue spruces? This species was very popular for Christmas trees until the nineties, and the ones released into the wild are now in some cases over twenty feet high. At this point they become a problem, because they're at risk of falling over in a storm and landing on the house. Either they have to be felled by a special company, or the owner ignores it and leaves the tree (and the problem) to grow bigger and bigger. Nordmann firs are in vogue today – so these are the ones to watch out for in thirty years' time when they're enormous and taking over our gardens …

What's it like for the young trees to come inside for Christmas? Well, they're dormant as they're mid-hibernation. Rather like hedgehogs or bears, they fall into a deep sleep, reducing their energy consumption to an absolute minimum. They save the reserves they built up last summer to form new shoots in the spring. Spruce and fir trees recognise when it's time to start growing by the temperature and the length of the day. If both warmer temperatures and more light come together, then it must be the start of spring – that's what millions of years of inherited experience tells them. Except that this isn't true in the living room. The festive lights are on until late in the evening, and the central heating or the wood burner keeps the room warm like the summer sun. Winter is over for the small fir trees, but only for a few days. By the middle of January at the latest, it is sent back out into the open, straight back into winter. Many trees cope valiantly with the challenge of having to switch back into winter mode. But for some trees, this moving back and forth between seasons is fatal.

24. Woodland Walks in February

Isn't February dire? It's a miserable month to be outside. Here in Germany, the trees are bare, it's cold and damp, but mostly not cold enough for snow, thanks to climate change. Instead, days of rain have softened the ground to a squelchy mire, so with every step mud splashes up your trouser legs. The long, tedious wait for spring has now reached its zenith and our mood is correspondingly low. But just because we've got the winter blues, it doesn't mean that out in the woods it's all doom and gloom. Once you pick yourself up and go out for a walk, you'll find that this supposedly barren season is by no means as boring and colourless as you might think. Quite the contrary.

First there's the moss. It grows on the underside of tree trunks and on the roots at the base of the trunk, giving the impression that green octopuses have conquered the forest floor, with trees sprouting up from amidst the tentacles. The contrast between brown foliage, grey-brown bark and bright green moss is especially vivid at this time of year. We might find a touch of white if it snows. Or even, given the right conditions, we might be lucky enough to see hair ice: a white, fluffy and rather magical phenomenon that appears on dead branches rotting on the forest floor. This curious sight is conjured up by fungus within the branches, which is decomposing the wood: as it does so, it digests and, like us, emits water vapour, carbon dioxide and other organic compounds. This 'exhaled breath' freezes as soon as it reaches the cold air outside. As every outbreath carries on freezing, it forms curling strands of ice like fine silver hair or candyfloss. If you touch it, it instantly melts into water droplets.

If the wood is frozen through, then the fungus can't function because it also freezes. Therefore, you'll only find hair ice when the air temperature is slightly below zero degrees but when the rotten wood on the ground is still above freezing.

Some trees and shrubs are already springing into life, such as the hazelnut. Its male flowers, or catkins, hang from the branches like lambs' tails and scatter pollen – kickstarting the hay fever season for unfortunate sufferers. While the deciduous trees are still dormant, the conifers are already getting ready for spring. In their original homeland, in the far north, they have to make use of every warm day in the short growing season, so they start much earlier than their broadleaf counter-parts. From the outside, there's barely any sign of activity, because the buds for the new shoots are still closed. However, if you ever pass a spot where some trees have just been felled, it is worth taking a look at the stumps. In warmer weather, the tree stump will still weep droplets of resin at the edge near the bark, which shows that the tree is still attempting to pump moisture into the wood. The tree still isn't dead though it's been felled. Fresh sap in the tree always marks the beginning of the new growing season. The pressure of the flowing sap continues to rise in March and April and won't be interrupted even by a brief cold spell with snow. That's why this is the harvest season for maple syrup. As soon as the leaves and fresh shoots emerge, the pressure in the trunk decreases again, and the wood becomes a little drier.

A thick blanket of slowly melting snow is ideal for trees, as the water can seep gradually into the soil and can be stored away long-term. Woodland trees can draw on these reserves in extended dry spells in the summer.

In February, native birds are becoming more active in their search for a mate and are increasingly vocal in defending their territories. Towards the end of the month you might well hear woodpeckers drumming. This is their way of singing and communicating to competitors that this patch of the woods is occupied. Hares start getting spring fever around now, sometimes even in January. The females are picky about their mate, and will see off any males who are overly persistent. If you look closely at the ground after a mad March hare boxing match, here and there you'll see clumps of plucked fur.

25. Woodland Walks in May

THAT TIME IS HERE: the deciduous forests are turning green again. In the highlands, the lyrics of the old German song about May ring true, with trees bursting forth with life; meanwhile, at the lower altitudes, climate change has already brought this phase forward into April. For the trees, it takes a tremendous feat of strength, which all but exhausts the stored reserves from the previous summer. So, not wanting to waste a drop of energy, they sit tight until they're sure it really is spring, waiting until the hard frost seems unlikely to return before they produce fresh buds. But even trees can be wrong, and sometimes in the hills you can get a frosty spell even in June. Then the fresh green foliage hangs limp and brown on the branches, and beeches and other deciduous trees are faced with a fierce struggle for survival. Everything has to start all over again, and not every tree has enough energy reserves to produce new shoots twice in a row.

Trees are particularly sensitive at this time of year, with gallons of water being pumped up through the trunk. A few weeks earlier, in March and April, the pressure is so high that you can even hear the water drawing up through the trunk if you hold a stethoscope against the bark. How the green giants fill themselves up with moisture has still not yet been definitively explained. Transpiration, osmosis, capillary action – none of these theories explain it fully. Due to the large quantity of water in the sapwood, the bark no longer sticks quite so firmly, making trees particularly vulnerable to injuries in the spring. And the humidity at this time of year means that any wounds risk becoming infected by fungus and bacteria in next to no time. This makes it especially difficult for wounds to heal, which is why garden trees shouldn't be pruned in the spring. If a branch is sawn off a deciduous tree in March or April, the tree will keep drawing water through, which drips from the wood as sap. The tree is bleeding, we say in German – *der Baum blutet* – and it's true.

In Germany, we have an official ban on pruning and felling trees from March onwards, however this is less for the sake of the plants than of birds: German legislators wanted to prevent disturbances during the breeding season. The law doesn't apply to forestry, however, which causes by far the greatest damage. Hundreds of thousands of nests fall victim to the timber harvest every year. Spruce and pine trees are felled and with them down fall the nests that are hidden away in the crowns. Such collateral damage is seen as an inevitable consequence in the timber market, where sawmills apparently need to be supplied 'just in time' to meet year-round demand.

In early May, the forest floor is covered in some places by a carpet of flowers. In our latitudes, a natural forest is generally too dark for flowers. Only 3 per cent of the sunlight breaks through the leafy crowns of beech and oaks to the forest floor – for most herbaceous plants, that's not enough. But there is a narrow window of opportunity in spring, when there's still a chance for the flowers. If it warms up by the end of March, the delicate shoots of wood anemones, lesser celandine or wild garlic will emerge amidst the dry foliage of the previous autumn. These so-called 'early bloomers' have to hurry. They need to bud, flower, seed and put away reserves for next spring, all before it gets too dark again on the woodland floor. The huge trees that tower over them are still asleep and only wake up very slowly at the end of April. It will be mid-May before the canopy finally closes over. These colourful plants have just under two months to do what other species can spend all summer on. In this sense, wood anemones and their companions are the sprinters of the forest.

The merry month of May is also when certain large insects emerge from the ground. These are the cock-chafers (known as *Maikäfer* – May bugs – in German) that until then have lived in the soil as grubs – thick, white larvae – for three to four years. There, to the dismay of foresters, they feed on the tree roots, until they finally pupate and spend the winter as mature beetles deep under-ground. Once they emerge, these flying insects carry on feeding on foliage in the treetops and can deplete entire swathes of a forest when there's a population boom. Fortunately, they don't tend to cause lasting harm to the trees, however, as the trees will produce leaves again once they're gone at the end of June.

May bugs, also known as doodlebugs, were considered rare for years, presumed to be in danger of extinction. In 1974, the German singer-songwriter Reinhard Mey even lamented the end of the cockchafers. Since then it has been observed that as well as having a four-year life cycle, which corresponds with the developmental period of the larvae, populations swell and decline according to a thirty- to forty-five-year cycle. Incidences of mass propagation occur at intervals of three or four decades, followed by population collapse due to disease. It was this that led to the belief that the insects were close to extinction. In the past, cockchafers were not only feared for their insatiable appetite for tree foliage, including the leaves of fruit trees among others; no, they were also popular as a delicacy. Even in the twentieth century, people ate them raw, fried or boiled as the key ingredient in cockchafer soup. Confectioners even sold the little protein bombs coated with sugar. Others of a more sensitive disposition at least made use of this manna from heaven as free chicken feed, as my father recalls.

Larger are the rare stag beetles, living secret and hidden inside rotten wood. The larvae spend a happy few years munching their way through crumbling wood: it can be three or even up to eight years before they pupate and emerge in the world with their imposing mini antlers. In their adult phase, stag beetles live for a mere few weeks, and that's only to mate and lay eggs. The males' distinctive 'antlers' have evolved from jaws, or mandibles, and are used only to wrestle with rival males. As alarming as he may look to the uninitiated, this proud little warrior isn't dangerous. He can't bite, he can just lick a little sap from the trees. The females (which can bite!) scratch small wounds in the bark with their more normal-sized mandibles to get the sap flowing. After mating, the female lays her eggs at the roots of dying or dead trees, and that's it – the parents' work on this earth is done. Since stag beetles rely on dead wood, they are a protected species at high risk of extinction; in commercial forests these days far too little space is given over for rotting oaks and other deciduous trees. However, there is a refuge which makes a good substitute: rotting wooden fence posts or the dead stumps of fruit trees. If you have either of these in your garden, leave it for the little guys.

These creatures are a good example of how biased our perspective is. If the larval stage accounts for up to 99 per cent of the stag beetle's lifespan, would it not make more sense to name it for that period of time? The crux of the matter is that we don't tend to see them for that lengthy period and only ever get a fleeting glimpse of the short-lived reproductive stage. This colours our under-standing of them and even leads to a misplaced sympathy, just as with the mayfly, also known as the 'dayfly' as it emerges from the water for just one day. It does get to spend that day having sex, mind you, but that's after over a year in its larval stage, which is spent in streams and ponds. We feel sorry for their short life, although for an insect they live to a comparatively old age in total.

Incidentally, swarms of insects can sound quite alarming – and I mean acoustically, not just the idea of them. I once experienced the noise that an infestation of caterpillars can make. An oak grove here at Hümmel was infested by a swarm of European leaf roller moth caterpillars. Millions of them were nibbling their way through the fresh leaves. And as we all know, if you eat a lot, you excrete a lot. The faeces of one single oak leaf roller caterpillar is a minuscule ball, but when there are tens of thousands of the little critters feasting on oak leaves, that's a lot of little balls falling continuously to the ground. The sound is reminiscent of heavy rainfall, except that it can be heard for weeks on end. It goes without saying that a stroll among the oak trees is rather less appealing than usual in these circumstances.

26. Woodland Walks in August

A STIFLING SUMMER HEAT HANGS OVER THE TREE TOPS, and it seems as though it's not only the hikers who are exhausted – so are the trees. Looks don't deceive us this time: beech, oak and the other deciduous trees are already slowly preparing to shut down for the winter. By means of photosynthesis, they have already filled up their sugar reserves under the bark and in the roots, so little remains for them to do other than hunker down and wait for next spring. Their disposable leaves – the single-use solar panels that are only good for one season – are already pretty worn out. Many will be marked by traces of invasive insects such as the beech leaf-mining weevil. The little rascals lay their eggs inside the leaves and the larvae eat their way through, making serpentine tunnels in the leaves known as 'mines'. These mined areas turn brown, giving a heavily infested tree more of an olive colouring from a distance than fresh green. The adult beetle carries on causing mayhem just as it did in its youth, taking bites out of the already battered leaves. The poor old beech finishes the summer with its leaves in tatters, shot through with holes, as though a tiny woodland sprite had used it as a firing range.

As we saw in the chapter 'Forest Survival Skills', the spruce cambium is only good to peel off and eat until the beginning of July, because at this point the trees gradually start to withdraw their sap from the tissue. It becomes hard and woody, and we see something similar with leaves and conifer needles. By August, they have lost their juicy green colour and are tinged instead with yellow; their former vitality has given way to a certain weary languor.

The summer heat and lack of rainfall sometimes amplify this effect. Drought can lead to many trees shedding some of their leaves mid-summer. The birch tree in our garden often does this in late July to keep the rest of its foliage green until October. Cherries and rowan berries have often soaked up enough sunshine by August to have formed all the sugar they need. Once the reserves are full, they can shut up shop and put their feet up for the winter. Their leaves turn red and their metabolism shifts onto the back burner until spring.

Even the birds seem to be flagging. At least, you hear a lot less of their cheerful song and the drumming of woodpeckers. Forest dwellers are generally a bit quieter, anyway. The stock dove has a limited range of calls. It is as big as a wood pigeon, but without the white patch on its neck. And instead of the wood pigeon's extended 'hoo-hoo-ooh-hoo-hoo', all you hear from the shy woodland stock dove is a restrained 'hoo'. But in August, even this delicate little call is no longer necessary, as the breeding season is over, and there's little more to show off about. The woodpeckers make much less of a racket with their hammering, for the same reason. Many forest birds raise chicks just once a year.

After all, their food supply of insects and fruit is strictly seasonal, and by late summer the bulk of what's on offer is already over.

Does that seem surprising? Many flowering plants are still in full splendour, and our gardens see a corresponding rush of insects. The brambles are also now overloaded with fruit; you'd have thought that would be enough to sustain later breeding. But this autumnal abundance is typical of open steppe landscapes, as represented by our meadows and shrubs. There, in the open landscapes that are synonymous with our origins as humans, summer is still in full bloom, while the shady forest is already preparing for winter. There's barely any sign now of the aphids, which back in the spring were sucking the sap of billions of fresh leaves and shoots. The larvae of beetles and flies have long since pupated into adulthood and in the lengthening shade of the trees, they're preparing for hibernation under loose bark or leaf litter on the ground – a dormant state for over-wintering technically called 'diapause'. Little wonder then that birds can't sustain another brood: the abundance of the forest is over and precious calories are scarce. This makes it comparatively quiet in the late summer woodlands, and I have often been asked when leading guided walks why there are so few birds in Hümmel Forest. Paradoxically, it is very different in clear-felled areas, where there is often more of a steppe ecology. There, where all the trees have been cleared, flowering plants spread, such as the magnificent perennials of foxgloves and rosebay willowherb. With their

metre-high red inflorescences, they attract bees and
bumblebees as well as other nectar-sucking insects. Here,
songbirds still find plenty of food and so may manage
up to three broods per season. And their singing season
is correspondingly long.

27. Woodland Walks in November

THE TREES HAVE SHED THEIR FOLIAGE, the sky is grey, and cold drips fall from the branches. Who likes going for a walk in this miserable weather? And yet, you appreciate the November drizzle more when you stop to think about what's happening among the trees. It's not for nothing that rain is called 'liquid sunshine' — after all, it's as vital to the forest as the sun's rays. In summer, in our part of Europe, it doesn't rain anything like enough, or rather, the trees consume too much. A fully grown beech absorbs up to 500 litres of water on a hot day, and even if there are heavy thunderstorms, the supplies that fall from the sky are far from sufficient for these thirsty giants. So, they have to put away reserves and this is what they do over the winter. So when it seems to be raining constantly, you may find it reassuring to think that the trees are tanking up for the summer ahead. There can be up to twenty-five cubic metres of water in the soil around the roots of a single tree, stored in the smallest pores. When modern timber harvesters irretrievably flatten these reservoirs with their enormous tyres and their weight of up to fifty tonnes, it's easy to see the consequences. The forest is full of puddles, which, apart from rare wetlands, is quite unnatural. The surplus left with nowhere to go after weeks of rainfall will eventually seep down to the water table, a process that can take many decades.

Incidentally, when the water does flow downwards, it is not just the pores of the soil it seeps away into. Below ground, there's an artificial irrigation system, built not by humans but by earthworms. As they tirelessly dig their way through the soil, they create a system of burrows lined with mucus. They can move relatively quickly through these routes underground, but often not fast enough, for on their tails are the moles grabbing at these juicy little morsels, which can be as thick as a pencil. If the moles catch more than they can eat, they immobilise their prey by biting them in half, then stow them away in their burrows as a living supply. Doesn't sound very nice, does it? And I'm quite sure it isn't for the worms.

Have you ever wondered why you see more earthworms after a spell of rain? If you don't like those gloomy autumn days when everything turns to mud, well, you're in good company, because earthworms also hate the rain. It floods their subterranean burrows, and if they don't squirm up for air fast enough, they face a wretched death by drowning. And that's not all. If one takes an unlucky turn and ends up in a puddle, it heads straight for the fate it wanted to escape. With the water struggling to flow through the compacted soil, it's the worms that feel it the worst; after a rain shower, hundreds of 'sailor's graves' can be found under the surface.

Incidentally, at this point we can once again make a brief detour back to emergency survival in the woods and meadows, because catching earthworms is a genuine alternative to hunting. Not only in rainy weather; no, even in the sunshine you can tempt them up to the surface by a method known in some parts of the world as worm charming or worm grunting. Simply tap a stick into the ground and start drumming your hands against it. This produces similar vibrations as raindrops, and after a few minutes the first worms will emerge. Alternatively, you could simply walk on the spot – tapping the ground is a method employed by several animal species to entice their worm prey to the surface. And then? Fried with a little salt, the taste is not unlike chicken. I've got nothing against them as a source of protein. The amount of worms may exceed one hundred tonnes per square kilometre. In our climes at least, even in times of crisis, nobody need starve to death.

Back to our November stroll. Autumn is also mushroom season. The first wave usually comes in late summer, when after a long drought, heavy raindrops finally moisten the forest floor. But those are the impatient ones who can't wait until the more typical start date in the autumn, with the first longer rainy spells. This approach is much safer for reproduction because the caps will last longer. Besides, just before their winter dormancy, trees have ample sugar supplies, which fungi can tap into to form sumptuous fruiting bodies. A delicacy not only for humans but also for wild boar. Most of all, these grey, hairy hustlers go mad for oily and starchy fruits, such as the seeds the oak and beech trees offer in abundant quantities in certain mast years – and that's when the feast starts for the deer too. These woodland mammals need the calories to build up winter fat quickly under the skin. When the winter cold sets in, their metabolism will switch down a few notches and they'll spend the days dozing in a sheltered spot, hidden from sight and from the elements.

Mice, squirrels and jays can be seen scurrying around in the autumn, hiding away their supplies in underground depots. While the jay can reliably find up to 10,000 of its stashes, the squirrel sometimes seems to have the odd memory lapse. Then, in spring, these forgotten caches become obvious as whole bunches of seedlings sprout from the abandoned seeds.

28. Forest School: Child-Friendly Adventures in the Woods

HOW ABOUT AN INTRODUCTION TO THE FOREST of a chewable variety? Have your kids ever had a go at making their own woodland chewing gum? This is something I learned about in Sweden when I was an apprentice in 1984. During a tour of the southern part of the country, we visited some forestry companies that showed off how efficient their operations were with their industrial machinery. The participants were given handfuls of information at each site, but one of the companies included a small flyer describing how to make spruce chewing gum. I have no idea what the company was attempting to achieve by including such information, but when I tried it I found the recipe worked perfectly; and it is something children would enjoy as the highlight of a forest walk.

First of all, you need to find a suitable spruce or a pine tree. This shouldn't pose too much difficulty as these tree species happen to be the most common in our commercially managed forests. A suitable tree is one where there is resin seeping out. Resin is the tree's sap, its lifeblood, which, like ours, always flows when the bark or skin is injured. Now, you shouldn't break the bark just to make yourself some chewing gum, and it also wouldn't help anyway, because you need to find a blob of resin that has already hardened on the bark of the tree. This hardened resin should be clear and at least the size of a fingernail. When you've found some, put it in your mouth and suck it to gradually warm it up. Every now and then try carefully to see if you can chew it yet. Don't bite too hard until you're sure it's softened up, as if you bite it when it's brittle the resin will crack into small fragments, and then you'll need to start again with a new piece. If the resin is clouded with white cracks, then it won't work and will disintegrate in your mouth into a bitter-tasting dust – not what you're after. Even if it works as intended and the resin gradually gets softer and more malleable on your back teeth, it will taste rather sour at first. You can spit this bitter juice out. I know, it doesn't sound very elegant, but you're in the woods, and nobody is watching except your group and a few birds. Gradually, the taste becomes more palatable and the resin turns into a pink forest chewing gum that doesn't stick to your teeth. It's a neat little surprise for when the kids are after some excitement.

But be warned, if you don't find the right kind of clear, hard resin, and you have a go with some that's cloudy or still sticky to touch, then I'm afraid you'll be frustrated and disappointed as it just sticks to your teeth and doesn't taste nice at all. When you've had enough of chewing, you can dispose of this natural product in the undergrowth or leave it where you found it – simply stick the gum back on the bark of the tree.

Children are grateful woodland explorers if you let them get stuck in. And that includes a bit of getting mucky. We adults often perceive dirt as repellent or even repulsive, and that's fine under normal circumstances. Much of the dirt we produce as humans such as oil, paint, soot, dust and of course pet excrement can harm us and needs to be quickly washed off if we get it on our clothes or hands. On the other hand, there's nothing unhealthy about a bit of mud or crumbly humus. Even the green algae film you find on tree bark (which is slippery in the rain and sticks to your jacket when you lean against the trunk) is harmless. It's not unlike seaweed.

Nevertheless, there tends to be an innate barrier that holds us back from intensive contact with nature, as I discovered when leading a group of challenging adolescents. With their white trainers and mobile phones, they were initially very reluctant about stepping foot in the woods. To avoid skidding as they walked gingerly between the trees, most found themselves a walking stick for support. But they didn't pick it up or hold it with bare hands. No, to my astonishment they pulled out paper tissues to touch the sticks with. After two days out and about in Hümmel Forest, I almost didn't recognise them. I had built some survival elements into their daily routine, and now the girls and boys were daring each other to eat longhorn beetle larvae.

So, the main thing is that children wear clothes they know they're allowed to get dirty, and then they're off. What about painting some tree faces, for example? All you need is a patch of bark on which you can get creative with some sloppy mud and a stick as your paintbrush. Soon the kids will be off, painting eyes, noses and mouths on the trunks until half the forest is full of funny characters. Who knows – they might last for a few days, perhaps even until the next walk?

What about making a forest telephone? It only works over limited distances but even over a short stretch it has an important role to play, at least for birds that nest high up in tree cavities. The most fearsome enemy for their chicks are squirrels or martens. These mammals climb up the trunk while the parents are out, and fish inside the nest hole with their sharply clawed front paws, trying to grab the helpless offspring. What can the parent birds do? Not a lot, but if they hear the predators coming they can at least launch a brave attack on the wing and try to scare off the assailants. Sometimes the predators also creep up on sleeping adults, which wake up and fly away just in time, if they hear or feel that danger is coming. Their early warning system comes in the form of the forest telephone, which in this case consists of the tree trunk. Wood conducts sound very well; that's why it's used to make musical instruments. Well, here the old trunk acts like a huge percussive instrument, where the squirrels and martens unwittingly tap out a song of death. As they climb up, the scratching sound of their claws on the bark travels up the wood into the nest hole. The chicks have just a few seconds to react to the approaching danger and screech for their parents' help.

Children can understand how the 'telephone' (or rather, the alarm system) works on old tree trunks lying on the ground. They kneel at one end and press their ear to the bark. At the other end you sit and tap out a message with a pebble, like Morse code. The children listen out and count how many times you tap. It's even more like the real-life scenario if you scratch the bark instead – then the children hear the same predator warning as the chicks.

The best parts of a hike are the breaks. Don't forget to stop for a snack, especially when you're out walking with children. I have found on many tours with primary school classes that it's best to stick to the routine of the normal school day. When we've got carried away with an experiment or a game and I've lost track of the time, the kids' attention quickly diminishes and that's when you start to hear the first grumbles. When everyone's appetite is satisfied, the children's enthusiasm picks up again and you can start another activity. How about a little woodland music? And I don't mean birdsong or the sound of the wind in the treetops. No, I'm talking about making some 'real' music. But mightn't this disturb the natural atmosphere? Don't worry, because the soundscape you can summon up with these instruments is nothing if not natural.

Let's start with the simplest instrument to master: a beech leaf. Place your thumbs together and you'll see there's just a small gap between your knuckles. Clamp the beech leaf here between your knuckles so that it's stretched tightly. That's it – your first forest instrument is ready! To play it, press your lips firmly against the gap between your thumbs and give a big blow. It makes a croaky squeak and it can be quite loud. By varying how hard you blow, you can vary the pitch and how scratchy it sounds. But that's about it for the musical possibilities.

Nevertheless, this whistling sound is made quite often in the woods, especially in the summer, by hunters as a means to attract randy roebucks (in German, luring deer in this way is called *Blatten* from the word *Blatt*, meaning 'leaf'). With practice, you can use this simple whistle to imitate the call of a doe in heat, keen to attract a mate. Younger bucks may react spontaneously, because this could be the first opportunity they've had, especially if an older male has long occupied the territory and doesn't tolerate any rivals. When they hear that call to action, sometimes a plucky young buck will try his luck, throwing caution to the wind. A hunter would use this opportunity to take a shot, but how much nicer to take the opportunity to quietly observe the animal from close up. For the beech leaf whistle to be effective, it is impor- tant to make just a few short signals and then take a break for a few minutes. If still no deer show up, you can try again with a bit more urgency. The aim is to arouse the desire of potential suitors who haven't quite made their mind up yet. For more tips, look online for instructions. As with any instrument, practice makes perfect.

The second instrument falls into the same category: the plastic watering can. This is another one you can use to tempt deer to play a duet with you. The key condition being, of course, that you're in an area with a red deer population. There may well be a map of your region which shows whether this is the case, as with this website for Germany: rothirsch.org.[34] Condition number two is the time of year: the rutting season, when these deer calls work, falls between September and mid-October at the latest. Red deer stags round up their harem, defending them against rivals and grunting from deep in their throats to assert their authority and make themselves heard. This hoarse call is called 'roaring' and that is indeed what it sounds like. This is where our watering can comes into play. The spout is the tube you blow into and the container itself makes a wonderful sound box for amplifying the sound. The best way to work out what the call is supposed to sound like is to hear it live – because if you are indeed in a red deer habitat, you may well hear a roaring stag in the distance. Put the spout to your lips and make a hoarse, rumbling call, as deep as you can. You have made yourself a decoy stag, and even if no one answers, at least the others having a walk in the woods (and their children) will enjoy the show.

Admittedly, these two instruments are not actually suitable for making music in the traditional sense. But in this regard, nature also has something to offer: you can easily whittle yourself a willow whistle. This is something I learned to do as a child, when we used to go on long hikes with our family friends. I remember being amazed at how you could fashion such a magical thing from something as simple as a stick. You need a penknife and a fresh willow branch with green bark or a new shoot of sycamore, a couple of centimetres in diameter. Cut off a stick about ten to fifteen centimetres long. It needs to be smooth without any twigs growing off it and no buds or bumps in the bark.

First, cut a shallow notch into the stick two or three centimetres from the end that will be your mouthpiece. Cut through the bark and into the wood: this will be the air hole on the top. Then cut a ring around the stick, roughly in the middle of the stick or three or four centimetres from the notch. Cut through the bark around the circumference, deep enough to reach the wood. Then gently peel off the bark on the mouthpiece end in such a way that it comes off as one tube. To do this, place it on a hard surface and gently tap on it on all sides with the knife handle to loosen the bark (not too hard, you don't want to break it). I remember my friends' dad muttering a rhyme as he tapped the bark loose: '*Ene, mene, miepe, die Saat, die ist riepe.*' So, if you want to carry on this tradition, you could sing something similar like 'Eeny, meeny, miny, moe, plant a seed and watch it grow' …

Once the bark is loose, you can twist it to pull it off. The best time to try making this willow whistle is early spring, when the bark is green and there's lots of moisture in the tree. But if the bark doesn't come away easily, you could suck the end to moisten it and then try tapping it again. Once you've slipped the bark off in one piece, make the notch on the top bigger, using your knife to cut out a chunk about two centimetres long. Then whittle a flat edge along the top of the exposed wood, from the air hole notch to the mouthpiece end. Next, gently slide the tube of bark back onto the stick as it was before, and you're done.

When you blow into your whistle, you can vary the pitch by sliding the bark tube back and forth. The result is amazing, and with a little practice you can play entire songs. It's a very simple yet satisfying object to whittle and the whole process is exciting for children. When you have a long walk ahead with the potential for whining, a little willow music is the perfect distraction.

Conclusion

THIS BOOK IS NOT INTENDED AS A REFERENCE BOOK, but as an appetiser. You're unlikely to remember everything I've told you, and neither do you need to. You might want to go back and look up sections when you come across something in the woods that prompts new questions. But more important than a book are the senses you carry around with you already: your eyes, your ears, your nose, your tongue and your sense of touch. Equipped with these, you have the perfect toolkit to go on enthralling expeditions into the woods near you. They are your forests – just waiting to be discovered.

Don't worry, we humans are no bother to the animal and plant world when we take a stroll in their habitat. Quite the contrary: we belong in this environment, too, as long as we go on foot and leave the natural world as we find it.

With this in mind, I hope you enjoy the large and the small wonders that you encounter in the wild, and if this guide whets your appetite for another walk in the woods or a forage in the forest, then it has served its purpose.

Notes

1. Carson, Rachel, *Der stumme Frühling* (*Silent Spring*), C.H. Beck, Munich, 1963, p. 296.
2. Krämer, Klara, Dipl. Biol., RWTH Aachen University, Institute for Environmental Research (Biology V), Chair of Environmental Biology and Chemodynamics (UBC), email correspondence of 30.03.2016.
3. Schmitt, Craig L. and Tatum, Michael L., *The Malheur National Forest: Location of the World's Largest Living Organism (The Humongous Fungus)*, United States Department of Agriculture, 2008, p. 4.
4. Hengherr, S. et al, in *Journal of Experimental Biology*, 2009; 212: 802–7; DOI: 101242/jeb.025973.
5. http://www.rki.de/SharedDocs/FAQ/FSME/Zecken/Zecken. html#FAQId3447426, viewed on 16.06.2016.
6. https://diepresse.com/home/panorama/oesterreich/5393071/ Viele-Kinder-nicht-geimpft-mehr-FSMEFaelle, viewed on 16.06.2016.
7. Fuhr, Eckhard, 'Bambi schützt vor Borreliose' ('Bambi protects us against Lyme disease'), in *Die Welt*, 25.05.2014, vol. 21, p. 2.
8. https://de.statista.com/statistik/daten/studie/226127/ umfrage/hektarertrag-von-getreide-in-deutschland-seit-1960/, viewed on 27.12.2016.
9. Request made in writing by Mrs Christine Kamm to Dr Christian Magerl of Alliance 90/The Greens on 04.03.2013; response from the Bavarian State Ministry of the Environment and Health on 29.04.2013; in print with ref. no. 16/16704 dated 03.06.2013.
10. http://www.bfs.de/DE/themen/ion/umwelt/lebensmittel/ radioaktivitaet-nahrung/radioaktivitaet-nahrung.html, viewed on 12.04.2016.
11. https://lua.rlp.de/de/presse/detail/news/detail/News/ kleiner-fuchsbandwurm-jeder-fuenfte-fuchs-im-land-befallen/, viewed on 27.12.2016.

12. Ibid.

13. Infectious Disease Epidemiology Annual Report 2014, Robert Koch Institut, Berlin, 2015, pp. 72–5.

14. http://de.statista.com/statistik/daten/studie/185/umfrage/todesfaelle-imstrassenverkehr/, viewed on 17.04.2016.

15. https://www.br.de/themen/ratgeber/inhalt/verbraucher-tipps/gewitter-blitz-blitzschlag-folgen100.html, viewed on 17.04.2016.

16. http://wildbio.wzw.tum.de/index.php?id=58, viewed on 19.04.2016.

17. http://isleroyalewolf.org/http%3A//www.cpc.ncep.noaa.gov/products/precip/CWlink/pna/nao.shtml, viewed on 14.07.2016.

18. http://www.yellowstonepark.com/wolf-reintroduction-changes-ecosystem/, viewed on 15.07.2016.

19. Beigang, T., 'Der Wolf bei den kleinen Mädchen in der Bushaltestelle' ('The wolf at the bus stop with the little girls') in *Nordkurier*, 18.08.2014.

20. BPOLD-B, 'Die Geschichte vom Wolfstransporter – alles nur Wolfsgeheul!' ('A tale of wolf smuggling – the boy who cried wolf'), press release from the German Federal Police, Berlin, 27.01.2014.

21. http://de.statista.com/themen/1199/strassen-in-deutschland/, viewed on 24.06.2016.

22. https://www.nabu.de/tiere-und-pflanzen/saeugetiere/wolf/wissen/15572.html, viewed on 24.06.2016.

23. Rothe, K., Tsokos, M., and Handrick, W. 'Animal and Human Bite Wounds' in *Deutsches Ärzteblatt International*, 2015; 112: 433–43; DOI: 103238/arztebl.2015.0433.

24. Bloch, Günther and Radinger, Elli H., *Der Wolf ist zurück* ('The wolf is back'), Bad Münstereifel/Wetzlar, 2015.

25. Drösser, Christoph, 'Glas unter der Lupe' ('Glass under the magnifying glass') in *Die Zeit*, vol. 39, 16.09.2004.

26. *Verkehr und Mobilität in Deutschland* ('Transport and mobility in Germany'), German Federal Ministry of Transport and Digital Infrastructure, November 2015, p. 6.

27. Puttonen, E. et al, 'Quantification of Overnight Movement of Birch (Betula pendula) Branches and Foliage with Short Interval Terrestrial Laser Scanning' in *Frontiers in Plant Science*,

29 February 2016; 7: 222; DOI: 103389/fpls.2016.00222. eCol-
lection 2016.

28. Huber, M., 'Forscher schauen 300 Bäumen beim Wachsen zu'
('Researchers watch 300 trees grow') in *Tierwelt*, vol. 23, pp.
24–5, 04.06.2015.

29. Moir, H. M., Jackson, J. C., Windmill, J. F. C.,'Extremely
High Frequency Sensitivity in a "Simple" Ear' in *Biology
Letters*, 9: 20130241.

30. https://www.welt.de/regionales/frankfurt/article127656776/
Immer-mehr-Wildkameras-erfassen-Spaziergaenger.html.

31. Ahnelt, P., 'Unterscheidung in blau und nicht-blau' ('Distin-
guishing between blue and not-blue') in *Revierkurier*, 2009; 3:
4–5.

32. Mantau, U., *Holzrohstoffbilanz Deutschland: Entwicklungen und
Szenarien des Holzaufkommens und der Holzverwendung ron 1987
bis 2015* ('The balance of raw materials in Germany: develop-
ments and scenarios in timber production and use' from 1987
to 2015), Hamburg, 2012, p. 65.

33. Füßler, Claudia, 'Der Baum der Bäume' ('The tree of trees') in
Badische Zeitung, 17.12.2016.

34. http://rothirsch.org/wp-content/uploads/2014/02/
RWVWaldD+Wald_140225.jpg.

16/7/14.